Airborne Weather Radar

A USER'S GUIDE

Airborne Weather Radar

A USER'S GUIDE

James C. Barr

IOWA STATE UNIVERSITY PRESS / AMES

Captain James C. Barr has logged over 12,000 hours in his eighteen years as a pilot. He has instructed senior midshipmen of the U.S. Naval Academy, has been a flight engineer line check airman, and also has a master's degree in business administration from DePaul University, Chicago.

The purpose of this book is to provide information on airborne weather radar operation and interpretation. The user of this information assumes all risk and liability arising from such use. Neither Iowa State University Press nor the author can take responsibility for the actual aircraft flying in severe weather or the safety of its occupants.

© 1993 Iowa State University Press, Ames, Iowa 50014
All rights reserved

Orders: 1-800-862-6657
Office: 1-515-292-0140
Fax: 1-515-292-3348
Web site: www.isupress.edu

Authorization to photocopy items for internal or personal use, or the internal or personal use of specific clients, is granted by Iowa State University Press, provided that the base fee of $.10 per copy is paid directly to the Copyright Clearance Center, 222 Rosewood Drive, Danvers, MA 01923. For those organizations that have been granted a photocopy license by CCC, a separate system of payments has been arranged. The fee code for users of the Transactional Reporting Service is 0-8138-1363-8/93 $.10.

♾ Printed on acid-free paper in the United States of America

First edition, 1993

Library of Congress Cataloging-in-Publication Data

Barr, James C.
    Airborne weather radar: a user's guide/James C. Barr.—1st ed.
        p.   cm.
    Includes bibliographical references (p.   ).
    ISBN 0-8138-1363-8 (alk. paper)
    1. Radar in aeronautics. 2. Radar meteorology. I. Title.
    TL696.R2B37   1993
    629.132′5214—dc20   93-17077

The last digit is the print number: 10 9 8 7 6 5 4

To my Uncle Stuart,

who gave me wings to fly,

and to my family,

Betsy, Mara, and Ilyssa,

who help keep my feet on the ground.

Contents

Preface

This book is written primarily for aviators who are confronted with circumnavigating severe weather using airborne weather radar. It is specifically targeted at pilots in major airlines, commuter airlines, corporate aviation, commercial aviation, governmental aviation training programs and ground school courses, and at private aircraft owners.

Very little has been written on airborne weather radar operation and interpretation. Even less information is available for practical application. My goal is to fill that void with a unique combination of technical data and real-life applications that pilots can immediately incorporate into their procedures. This book is the link between the engineer and the pilot.

The text provides technical radar data on a user-friendly basis and recommended procedures for accurately assessing severe weather areas. First, it introduces basic airborne weather radar theory and principles. Second, it provides a thorough discussion of all system components and pilot-operated controls. The principle of reflectivity and reflectivity correlations with rainfall rates, hail, turbulence, and thunderstorm intensities are explained. Attenuation, a concept critical to survival, is covered in detail. Third, the book extensively discusses the weather radar antenna beam and how its controlled manipulation, interrelated with other features of the radar, can provide the pilot with enough information to make accurate interpretations of weather systems. Additionally, an operational summary and avoidance strategies are reviewed. Fourth, the text covers pertinent meteorological data to be used in concert with airborne weather radar. A comprehensive framework for pilot weather briefing is provided. Finally, a mock flight is narrated incorporating all topics discussed.

To gain the maximum benefit from this book, it must be reviewed regularly. In order to readily apply this information the pilot must understand it thoroughly, internalize it, and practice it regularly. Even on a clear day when weather is not a factor, the pilot should experiment with tilt management principles.

The dynamics of thunderstorms make severe-weather flying particularly demanding. To ease the burden of dealing with this ubiquitous phenomenon, well-informed decisions are vital. The intent of this book is to provide data to improve your decision making. It is my hope that it will make your life much easier and, more important, much safer.

Above all, know your own personal limitations – these are the parameters within which *you* must operate. Don't do anything you don't feel comfortable doing.

Airborne Weather Radar

A USER'S GUIDE

1

Radar Theory, Components, and Hardware

Radar Theory

The word "radar" is an acronym for RAdio Detection And Ranging. The concept was devised by Heinrich Hertz. He discovered that if you transmitted a radio wave, the signal would travel out until striking a target. The signal would then reflect back, or *return,* to the transmitting unit. This return can be thought of, and is often referred to, as an *echo.* The speed of this radio pulse is approximately 300 million meters per second, almost the speed of light. Therefore, the distance of an object that is struck by the energy wave can be determined by the equation

$$D = V \times T \qquad \text{where} \quad \begin{aligned} V &= \text{velocity} \\ T &= \text{time} \\ D &= \text{distance} \end{aligned}$$

In the case of airborne weather radar, a transmitter emits a microwave pulse of 8,000 to 12,500 megahertz (MHz), which is distributed by the antenna. The microwave continues out until it dissipates or strikes an object, at which point it is reflected back

3

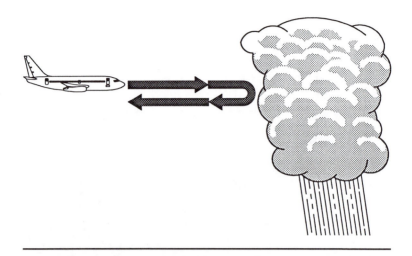

Fig. 1.1 A simplified drawing of the energy flow from and to the radar unit.

to and collected by the antenna. It is then directed to the receiver, processed, and depicted on the *plan position indicator* (PPI). The radar, by itself, cannot distinguish between ground returns and weather returns.

This obviously is a succinct description of radar theory. However, we will expand on the theory as the book progresses.

Radar Components

All radar systems have three basic components:

1. Receiver/transmitter (R/T)
2. Antenna, of which there are two types: parabolic and phased array (flat plate)
3. Plan position indicator (PPI) and indicator control panel

Fig. 1.2 The basic components in every radar system: *a*, parabolic; *b*, flat plate (phased array). (Courtesy of Allied-Signal Air Transport Avionics.)

Receiver/Transmitter (R/T)

Within the receiver/transmitter (R/T) unit is a component called a magnetron. It is a vacuum tube responsible for producing a high-powered pulse of radio noise. This high radio-frequency (RF) noise is a microwave pulse/energy traveling at a specific frequency. The use of high-powered RF energy enables a smaller antenna to be used. Without the magnetron, antenna size would be increased so much it would be prohibitive for practical use on aircraft.

The R/T unit transmits the microwave produced by the magnetron. After the transmission has occurred, the transmitter shuts off and the receiver turns on. The receiver then listens for "echoes" generated by the previous transmission of the microwave. If any echoes are detected, they will be depicted on the plan position indicator (PPI).

The R/T automatically and continuously alternates between transmission and reception. Each time this sequence is completed, the antenna travels another fraction of a degree and the R/T cycle is repeated. This cycle is known as the *pulse repetition frequency* (PRF) and on average occurs approximately 200 to 1,500 times a second, depending on the unit. For example, if the antenna scans at a rate of 150 degrees per second and the PRF is 1,500 cycles per second, then each degree would be allocated approximately 10 transmissions and receptions (1,500 divided by 150 = 10). Therefore, although it may appear on the PPI as if the antenna is making one continuous sweep, the picture is actually the result of a series of individual transmissions and receptions. The fact that this all occurs at close to the speed of light gives the illusion of one continuous sweep. This horizontal sweep is also called the *azimuth scan*. The component responsible for the distribution and accumulation of the microwave signal is, of course, the antenna.

Antennas

Antennas come in essentially two types: (1) parabolic and (2) phased array, also referred to as flat plate. Both types are commonly used, but the flat plate, for reasons discussed later, is clearly a more effective antenna. However, for our purposes it may be more beneficial to envision the parabolic dish type when introducing the antenna.

You might remember that a long time ago, when you were taking your math class in calculus or trigonometry, the teacher discussed "lines tangent to a parabola" and formulas computing the "focus of a parabola." As students we always wondered what useful purpose this could possibly have; well, lo and behold, these principles are at work in our weather radar antenna.

The microwave energy travels through a device called a waveguide (a circular or rectangular length of tubing), which "guides" the microwave to the antenna assembly. From the waveguide the transmission travels through the feed horn, which is located on the main axis of the parabolic dish. The nozzle-type mechanism attached to the end of the feed horn is the *splash plate*, or *subreflector*, which is located at the *focus* of the parabolic dish.

The splash plate distributes the energy back onto the parabolic dish, which reflects this energy forward in a beam (parallel to the main axis) shaped like a cone, or pencil point, similar to the shape of a flashlight beam. If any microwave energy strikes a target, the energy is reflected back and collected by the antenna dish.

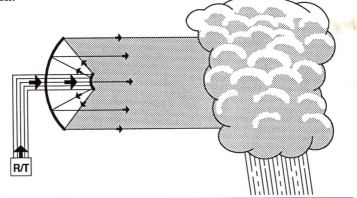

Fig. 1.3 The energy flow between the antenna and target is an alternating one-way street. Here, in the transmission phase, energy flows from the antenna to the target.

The principles of a parabola state that energy striking the inside of a parabola will be reflected back to the focus—in this case, the splash plate. The signal then travels back *in*, the same way it came out, and is received by the R/T. Any targets are then depicted on the PPI.

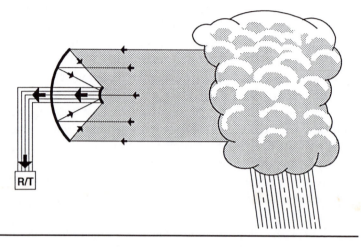

Fig. 1.4   Here is the reverse flow of energy during reception. Remember, the R/T can transmit and receive but can do only one at a time.

The energy radiated from the antenna is composed of a *main lobe* (where the bulk of the energy is transmitted) and smaller *side lobes.* Collectively these lobes form the *azimuth pattern* of the antenna beam. Side lobe losses are the source of false targets and picture distortion. Flat plate (phased array) antennas are more efficient because they concentrate more of the energy in a well-focused conical beam and consequently do not incur as much side lobe loss as a parabolic antenna. Since the beam is more focused, flat plate antennas require a greater degree of proficiency in tilt management but offer the capability to achieve a more accurate picture.

In an effort to further refine the beam generated from the antenna, engineers have attempted to design both systems (parabolic and flat plate) so that we, as pilots, only see the returns

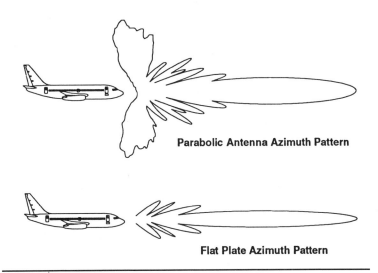

Parabolic Antenna Azimuth Pattern

Flat Plate Azimuth Pattern

Fig. 1.5 Typical azimuth patterns of both the parabolic and flat plate antennas. The flat plate antenna offers a more focused beam with fewer side lobe losses.

generated by the conical beam. However, side lobe losses still occur.

The angle of the conical beam is a function of radar frequency and antenna size. The radar frequency is in the *X band* (the operating X band range is approximately 8,000 to 12,500 MHz). The antenna size on your aircraft, of course, is of fixed size. The larger the antenna, the smaller the angle of the conical beam. In turn, the smaller-angle beam facilitates a higher concentration of energy. Most air carrier operations use a 30-inch antenna, which produces a 3-degree radar beam. Table 1.1 illustrates various-sized antennas and their corresponding beam angles.

Table 1.1. Beam angles of various-sized antennas

30-inch antenna	=	3-degree beam
18-inch antenna	=	5-degree beam
15-inch antenna	=	7-degree beam
12-inch antenna	=	8-degree beam
10-inch antenna	=	10-degree beam

Note the inverse relationship between antenna size and the angle of the beam. The smaller the beam angle, the more concentrated the energy. The smaller the beam angle, the more accurate the portrayal of the weather.

The 3-degree beam is one of the more narrow conical beam dimensions and enables a target, such as a thunderstorm (TRW), to more easily *fill* the beam and thus provide a more accurate picture on the radar scope (PPI). This concept of *filling the beam* involves *beam widths* and *beam width calculations* and will be discussed extensively in Chapter 3.

An important aspect of antenna operation is antenna stabilization, and I will discuss this feature of the antenna in some depth.

ANTENNA STABILIZATION

The purpose of antenna stabilization is to maintain a constant antenna sweep relative to the horizon during *moderate* aircraft maneuvers. That is to say, if a pilot has selected a particular tilt for the radar antenna after evaluating the weather and decides to deviate around an area of TRWs or decides to turn, climb, descend, accelerate, or decelerate for any reason, the antenna will maintain the sweep and the tilt selected by the pilot. This provides the pilot with a constant and, more important, a useful picture on the PPI. If the stabilization is turned off or inoperative, the antenna is locked in the attitude of the aircraft and, consequently, the antenna scan mimics the aircraft's maneuvers. Therefore, stabilization is a "good thing." Although it is seldom discussed, it is *the* mechanism maintaining an accurate picture for us as we maneuver our aircraft around TRWs.

To facilitate antenna stabilization, the antenna is referenced to the aircraft's vertical gyro. The antenna receives roll and pitch information from the vertical gyro and then compensates antenna movement so that the scan remains in the same place you put it. Unfortunately, the system has some limitations, most of which are due to precession. They are (1) pitch errors, due to acceleration and deceleration in takeoff (T/O) and landing (LND); (2) turn errors, occurring after takeoff during turns; (3) errors due to shallow banked turns; and (4) errors due to exceeding the excursion limit, which is determined by mechanical and electronic factors.

Let us take the last drawback first. Although the antenna stabilization will compensate for many aircraft maneuvers, it does have limits; that is why I said to maintain stabilization during *moderate* aircraft maneuvers. If you whip into a 45-degree bank, don't expect the stabilization to keep up with you; it cannot. In fact, stabilization capabilities regarding excursion limits are a function of pitch, roll, and antenna tilt. If the sum of these inputs exceeds a certain value, the antenna stabilization capabilities will be exceeded. (Commonly used limits are 25 or 43 degrees, but check your radar owners manual for *your* specifications.) For example, if your excursion limit is 43 degrees and you have 40 degrees of bank and 9 degrees of tilt, your radar's mechanical stabilization limits have been exceeded by 6 degrees (49 − 43 = 6).

Stabilization errors can also occur during shallow banked turns, that is, turns of 6 degrees or less. When the aircraft is banked 6 degrees or less, mechanisms called *erection torquers* provide false information to the vertical gyro as they seek a false vertical or false wings-level attitude. You may have already observed this error when intercepting a radial with a shallow cut and shallow bank. After you roll wings level, the attitude indicator still shows some degree of bank. This erroneous information is translated into a *stabilization error*. It will last about as long as you were in the turn. This type of error does not occur at bank angles greater than 6 degrees because then the erection torquers system is automatically disconnected.

Pitch errors and turn errors can be addressed concurrently. Pitch errors occur during acceleration and deceleration, as in takeoff and landing. Once again, the erection torquers (during acceleration and deceleration) seek a false vertical, and this erroneous information is transmitted to the antenna stabilization. This error begins to diminish as soon as your speed stabilizes and will be completely eliminated in about 3 minutes. However, in the meantime, the pilot can compensate by increasing or decreasing antenna tilt for an increase or decrease, respectively, in speed. If you happen to be turning while accelerating or decelerating (or, more accurately, while pitch error is present), the turn will be translated into a turn error (or roll error) in stabilization. You now have an antenna that is temporarily giving a lopsided sweep, or what is called an *asymmetric return.*

So how can you compensate for and/or minimize stabilization error?

1. Obtain an accurate weather picture during your weather radar *confidence check* prior to takeoff. A confidence check simply uses the weather radar to interpret TRWs before you are airborne. It tells you whether your weather (Wx) radar is operational and reliable. (There will be a more lengthy discussion of confidence checks in the confidence check section in Chapter 3.)

2. Execute definitive turns (i.e., turns greater than 6 degrees of bank) when navigating around weather.

3. Compensate antenna tilt for gyro precession so you do not overscan or underscan weather targets. (This technique will be discussed in Chapter 3.)

4. Plan wider clearance around TRWs.

5. Realize the stabilization limits of your aircraft. You may be able to mitigate some of the stabilization error through well-informed tilt management (discussed later). On the other hand, excursion capabilities are fixed and, if exceeded, cannot be compensated for.

We are now ready to discuss the third component of our weather radar system: the PPI and indicator control panel. For the rest of this chapter, the focus of discussion will be on these items, which are directly manipulable by the pilot. A broad-based knowledge of their operation will enable pilots to "tune in," so to speak, the best weather picture available.

Plan Position Indicator (PPI)

The plan position indicator (PPI), sometimes called the scope, is the radar screen. It is the *cathode ray tube* (CRT) where our weather will be displayed. The PPI is set up to display range marks, which define distance, and azimuth lines, which define direction, referenced to the nose of the aircraft. The azimuth lines never change; however, various selections for the distance markers are available.

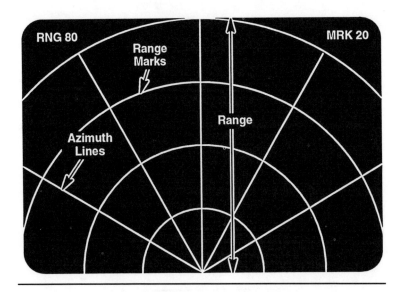

Fig. 1.6 Range, range marks, and azimuth lines are identified on the PPI. In this case, the selected range is 80 nautical miles (NM). Each range mark represents 20 NM.

Most antenna scans sweep 60 to 90 degrees either side of the centerline. Therefore, a full sweep comprises an arc of 120 to 180 degrees. (See your owners operating manual for exact specifications.)

Indicator Control Panel

The control panel of most radar systems has seven operating controls. They are (1) the function switch; (2) antenna stabilization on/off switch; (3) the gain control; (4) the antenna tilt; (5) the trace adjustment; (6) the picture erase rate knob; and (7) the range selector. (Trace adjustment and erase knobs are not normally installed on color radars.) Many radar displays will enunciate alphanumeric cues, in the legend of the screen, as various control modes/inputs are selected. Antenna tilt will be covered in Chapter 3.

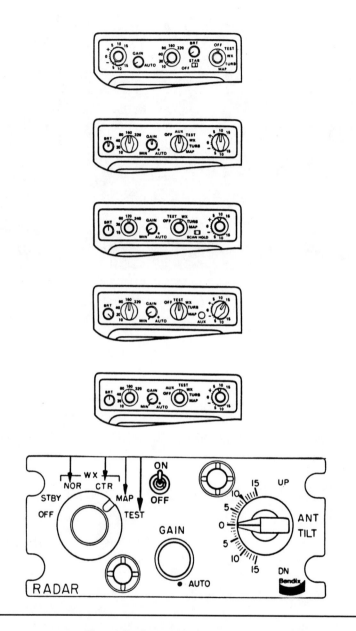

Fig. 1.7 Control panel layouts may vary, but the use of the function switches is generally the same. (Courtesy of Allied-Signal Air Transport Avionics.)

FUNCTION SWITCH

The *function* switch may have a standby mode. If so, it is important to operate in standby rather than off, unless you are extremely confident no weather will be encountered on your flight. This is recommended because when moving directly to a weather mode from the off-position a "system time sequencing" is engaged and a 3-minute warm-up (delay) is applied before any useful information is depicted on the PPI. If you inadvertently encounter an embedded rain shower while operating your weather radar in off, the 3-minute warm-up period will seem like an eternity as your aircraft proceeds ahead into what is now a known area of weather.

The *normal* and *contour* modes allow both weather targets and ground returns to be depicted on the PPI. Remember, your radar does not know the difference between a weather return and a ground return. Fortunately, however, the pilot can make that determination, primarily from manipulation of antenna tilt (discussed in Chapter 3). The contour mode is distinguished from the normal (or weather) mode in that it energizes an "iso echo circuit." As we have previously discussed, the radar transmits an energy pulse and listens, so to speak, for an echo, in the true sense of the word – that is, energy that is reflected back from striking a target. Of course, a more substantial target will reflect back more energy. For example, a heavy rain shower will reflect back more energy than an area of light precipitation. When contour is selected, the energy received from an echo is measured. If a substantial amount of energy is received back (i.e., 40 decibels in meteorological reflectivity, dBZ), the iso echo circuit will *not* illuminate the area responsible for the strong return, and therefore that area will appear as a dark or black hole on the PPI. A storm that takes on this appearance is said to be *contouring.* The contouring hole will be outlined in green, which represents the lighter area of rain. If your PPI offers a multicolored display, the contouring area will be red outlined by yellow and green (see Fig. 1.9).

Thus, the contour mode offers the pilot not only a two-dimensional display of length and width but also a third dimension in the form of rainfall intensity. The distance measured inward from the outer edge of green to where the storm begins to contour is referred to as the *gradient.* If this distance is short, the storm is said to have a steep gradient and is generally more well defined and stronger. A longer distance from the outer cell edge to where contouring begins is representative of a shallow gradient and thus a less well defined and most likely less severe storm than the one with a steep gradient.

An area of weather that is solid enough to contour is indicative of *at least* a level 3 "strong" thunderstorm, which produces a rainfall rate of 1/2 inch per hour or *greater*. Precipitation in this area would be from heavy to extreme. When confronted with a

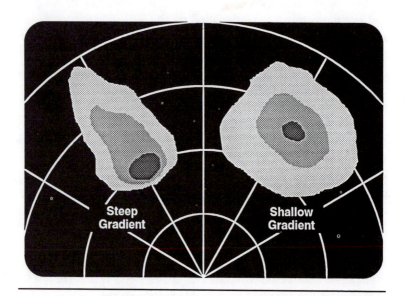

Fig. 1.8 The target on the left displays variable gradients around the perimeter of the return. The steepest gradient in this case is located in that part of the storm closest to the aircraft. This is the area where the *rainfall rate* is changing very rapidly with respect to distance and thus is suggestive of severe weather. The target on the right portrays a storm with a shallow gradient. The rainfall rate, when measured from the outer perimeter to the center of the thunderstorm, changes more slowly. When a shallow gradient is present throughout the storm, it is suggestive of less severe weather.

contouring cell, you have no way of determining whether it is a level 3 or level 6 storm. The important point, however, is that it really does not matter—a contouring area should never be penetrated.

When you are operating in normal, don't confuse isolated areas of black as contouring areas. Sometimes, they will be areas where no return at all is being received. In these cases you are most likely painting a valley between mountains, a river, or a lake. When painting a river or lake, the absence of signal is due to the energy ricocheting off the smooth surface and never making it back to the antenna. Shadowing, a form of attenuation, also displays a dark area, where no return is being received, and will be discussed in Chapter 2.

When approaching a weather system or deviating around weather, the radar should be primarily operated in the normal (or weather) mode with the contour mode selected intermittently just

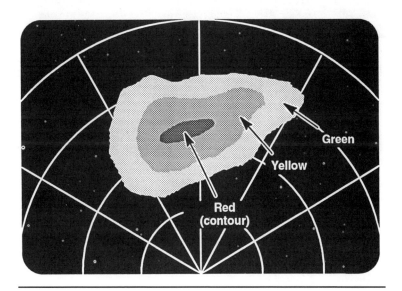

Fig. 1.9 In this black and white computer-generated simulation of a storm displayed on color radar, red is depicted as the darkest color; yellow surrounds the red and is represented by a somewhat lighter color; green surrounds the yellow and is represented by the lightest color.

to reidentify the areas of heaviest precipitation. Operating the radar continuously in the contour mode while flying in the vicinity of contouring weather may mistakenly lead the pilot into what appears to be a clear area but is, in fact, a contouring cell where the heaviest precipitation lies.

Digital radars offer color enhancement that makes this misinterpretation less likely. On these systems, contouring weather is displayed in red. Weather on the perimeter of the red will be displayed in yellow and/or green. Yellow areas are indicative of precipitation with less intensity than red but greater than green.

Note also that once an area of weather begins to contour, the entire weather signature should be avoided. The maximum level of turbulence may not be isolated to the contouring cell but may pervade the storm system.

Some digital radars have a *turbulence* mode function, which offers the pilot a fourth dimension of the storm: identification of

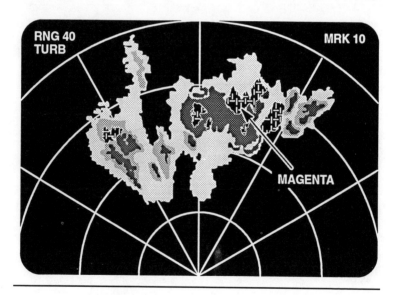

Fig. 1.10 In this display the turbulence mode has been selected. Not only is turbulence indicated in the contouring area of the storm (which would be expected) but it also is displayed in an area that is *not* within the area of heaviest precipitation.

areas of turbulence. When the turbulence mode is selected, a designated color (normally magenta) will be overlaid on the existing weather display. If no turbulence is displayed while operating in the turbulence mode, do *not* penetrate weather which would otherwise be avoided.

Since the focus of this book is on weather radar operation and target analysis, little will be said concerning radar ground mapping. Most radars have a map feature on the function switch. When this mode is selected, a slightly better picture of prominent terrain features is depicted on the PPI.

The *test* pattern on the radar should be compared with that suggested by the manufacturer's owners manual. In general, though, each feature of the radar will be incorporated into the system test pattern. For example, all the colors will be displayed, including the color designated for turbulence (generally magenta). On an analog (monochromatic) radar with a contouring feature available, a contouring band will appear somewhere in the test pattern. Essentially, the test feature offers a means to determine if all the video/pictorial capabilities of the radar are operational.

ANTENNA STABILIZATION ON/OFF SWITCH

The *stabilization* on/off switch is just that: in the on-position the antenna is stabilized with respect to the earth's surface (if the stabilization limits discussed earlier are not exceeded). The off-position will slave the antenna to the aircraft attitude. For example, if the aircraft was wings level and 0 degrees pitch, the antenna's sweep would be parallel to the earth's surface. If this attitude produced symmetric ground returns at the 60-mile mark and the aircraft rolled into a left bank, the antenna would now paint a lopsided (or asymmetrical) picture. Ground returns would now be indicated on the left side of the screen at a distance closer than 60 miles, and on the right side of the screen at a distance in excess of 60 miles, or, more likely, there would be no returns at all on the right side of the screen. This is because the antenna sweep is now canted to the left and no longer moving parallel to the horizon. The antenna sweep is now a function of aircraft pitch and bank.

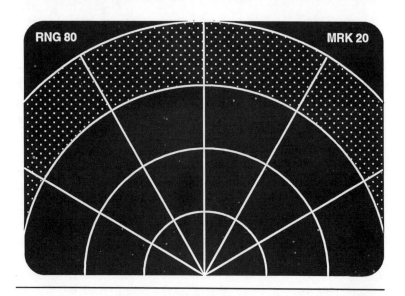

Fig. 1.11 In the diagram the aircraft is straight and level with the stabilization switch in the off-position. Ground returns are displayed beginning at 60 NM and are symmetric across the 60-NM arc.

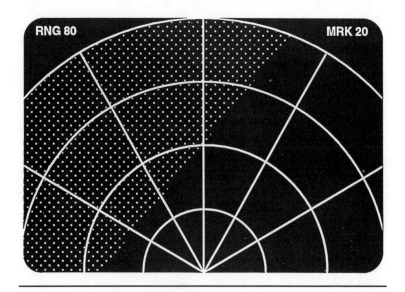

Fig. 1.12 The aircraft in Fig. 1.11, which is operating with the stabilization off, rolls left. The resultant display on the PPI shows asymmetric ground returns.

The only time the stabilization should be turned off is when the stabilization is inoperative.

GAIN CONTROL

The *gain* control provides a means to manipulate the "listening" sensitivity of the receiver. As you recall, the R/T transmits an energy pulse. The transmitter is then turned off and the receiver turned on to listen for the echo. If the gain is reduced, the receiver's sensitivity will be reduced, and it will pick up or detect only the loudest returns, or in our case, the areas of heaviest precipitation. In fact, the gain can be reduced to such a degree that it won't pick up any signal. Thus the gain control can be thought of as a volume control. In fact, the gain knob is set up in much the same way as a volume control. In *auto* the receiver's listening sensitivity is automatically calibrated. To reduce gain, turn the knob *counterclockwise*, or the same direction you would turn most volume controls to decrease the volume. As the gain is reduced, the entire display on the PPI will be affected. Areas of light precipitation and weak returns will diminish and eventually drop out completely. Areas of heavy returns will begin to change shape and appear smaller. At some point while the gain is being reduced the green area of a contouring cell will disappear and the contouring (black hole or red area) will change to green (if you are operating in the contouring mode). If using a color PPI, the red area will change to yellow or green. This is occurring because the receiver is no longer measuring 40 dBZ in return (the number is not important) but something less. Therefore, because of the programming in the iso echo circuit, the display that was contouring now appears only green (yellow or green on color screens). Remember, the requirements for contouring (a measured 40 dBZ in return) are no longer being met.

If confronted with several areas of weather, or if the PPI is saturated with returns, decreasing the gain will allow the softer (weaker) areas of precipitation to either fragment or drop out completely, while the most reflective cells will remain visible and well defined in shape. Thus the gain control allows the pilot to view numerous areas of weather on a relative or comparative basis. With the gain control in auto, some weather patterns may

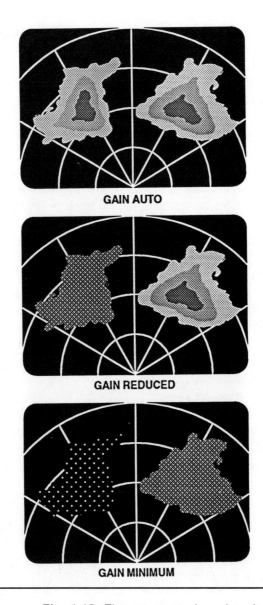

GAIN AUTO

GAIN REDUCED

GAIN MINIMUM

Fig. 1.13 These two targets change intensity at different rates as the gain is reduced. The target on the left appears to be fragmenting and decreasing in intensity more rapidly than the target on the right. This suggests the storm on the left is less severe than the storm on the right. However, both of these targets should be considered hazardous.

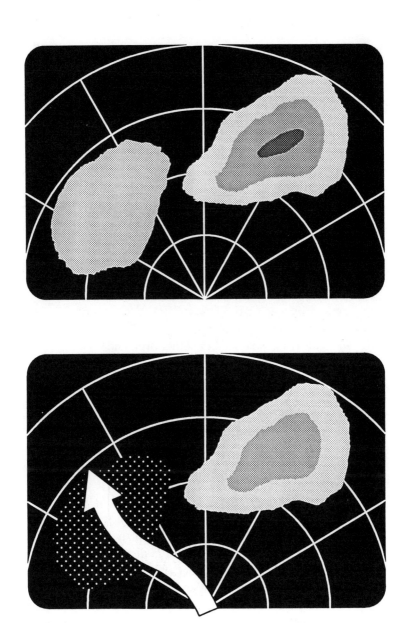

Fig. 1.14 In this particular case, reducing the gain has identified a possible deviation path that might not otherwise be revealed. However, in a scenario of this type one cannot overemphasize the need for extreme caution.

23

offer no clear-cut path of avoidance; reducing the gain may offer some insight as to which way to deviate by identifying the stronger areas of precipitation. Don't misconstrue this statement to imply that the gain be used to locate a path to *penetrate* a storm area. The entire thrust of this text is to focus on being informed enough to avoid getting into a situation like that. Explicitly, the gain control is only one of many tools the pilot can use to fine-tune the display on the PPI and augment information to make a well-informed decision.

As useful as the gain is, there are some cautions to be aware of:

1. When operating with the gain out of auto, the true strength of the storm will always be *understated*.
2. It is possible to decrease the gain enough so that all targets disappear.
3. Always return to auto before
   a. evaluating weather in another area and
   b. scanning for distant targets.

ERASE RATE AND TRACE ADJUST

The erase rate and trace adjust will be discussed concurrently since these two features are set at the same time, and normally only once. These adjustments should be made while testing the radar. After that, no further adjustments should be made during normal radar operation. (Note: Most color radars have a full-time weather/ground/test display and do not have either the erase rate or the trace adjust feature.)

The *erase rate* feature is almost self-explanatory. As the antenna sweeps across the sky, a picture is displayed on the PPI. The duration of that display is a function of the erase rate. As the erase rate is increased, the duration of the picture is shortened; as the rate is decreased, the duration is increased. The ideal setting is for the test pattern or target to be almost completely faded away when it is reilluminated by another sweep of the antenna. If the erase rate is too low (i.e., picture does not fade away or fades too slowly), the test pattern or the weather targets tend to *bloom*. (Blooming is manifested as a high-intensity bril-

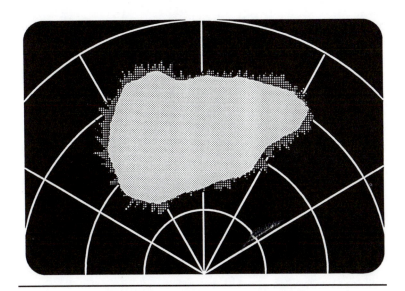

Fig. 1.15 Blooming produces a decrease in the visual sharpness of a picture, making it difficult to identify the perimeter of a return.

liance that flares up and causes the target to lack visual sharpness on the PPI.)

The *trace adjust* modifies picture intensity. A properly set trace adjust feature displays a test pattern that closely resembles that of the test pattern recommended by the manufacturer. If the trace is set too high, the pattern or radar returns will bloom. Worse yet, if the trace is set too low, known areas of weather will be understated or not depicted at all.

When the PPI displays a test pattern equivalent to that recommended in the manual *and* this picture is almost completely faded away when it is reilluminated by another antenna sweep, then both the trace adjust and the erase rate are properly set.

RANGE

Normally, three to five *range* selections (with their appropriate mark settings) will be available. Range equals the distance from the nose of the aircraft to the outermost arc on the PPI. Each arc is referred to as a *mark.* The range divided by the number of marks equals the incremental value of each mark. For example, a range of 80 nautical miles (NM) with 4 arcs implies each arc is representative of 20 NM (80/4). Therefore, the third mark out equals 60 NM (3 × 20). Thus any target on the third mark is 60 NM away from the aircraft.

Pictures on the PPI are very similar to television pictures in concept. A television picture is composed of a finite number of dots. This is also true of the PPI. The target you are painting consists of a number of little dots that are placed on the screen by an antenna sweep line. The dots are illuminated on the screen. For example, let's freeze a sweep line. We'll say a sweep line has 300 dots (this number is fixed). On a 30-NM range a dot is illuminated every 0.1 NM (300/30 = 10 dots per 1 NM, or viewed another way, 1 dot at each 0.1 NM). However, on the 180-NM scale, a dot is illuminated approximately every .6 NM (300/180 = 1.6 dots per 1 NM).

In the upper diagram of Figure 1.17 (range = 30 NM), the picture resolution on the PPI is much higher than in the lower diagram of the same figure. Therefore, targets displayed on the shorter range will appear better defined and, in fact, will portray the target more accurately, whereas depictions of the same target at the same distance viewed on longer ranges tend to appear "fuzzy" and a bit more nebulous.

Don't be deceived into thinking that the beam changes shape when you change ranges—it does not—or that the target is in some mysterious way moving closer and farther away as the various ranges are selected—it is not. If an area of weather is 100 NM away and you are operating in the 80-NM range, the weather will not be depicted. Conversely, if an area of weather is 50 NM away, it will be depicted on both the 180-NM range and the 80-NM range, but the 80-NM range will provide a sharper and clearer picture.

Good operating technique suggests frequent changes in the range control when aviating in the area of thunderstorms or rain showers. A very general guideline to follow is that long, medium, and short ranges should be correlated with long-, medium-, and short-range planning, respectively. The long-range scale provides

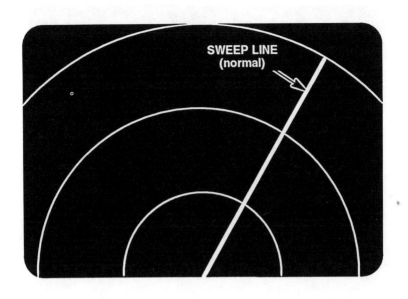

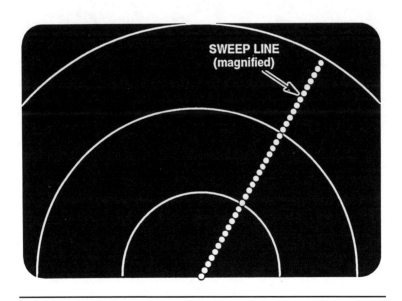

Fig. 1.16 A simulated sweep line's appearance when viewed normally and when magnified.

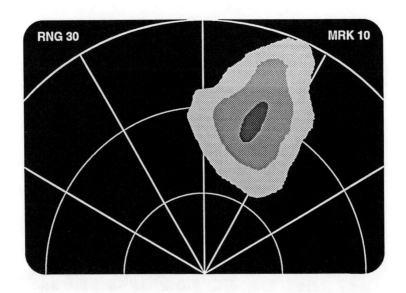

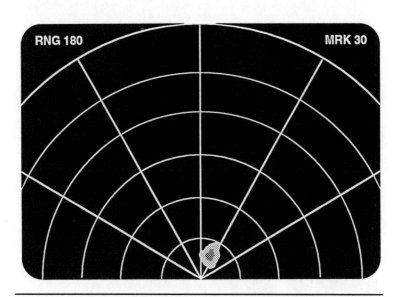

Fig. 1.17 Here is the same target at the same distance when viewed using two different range settings. The shorter range scale offers a higher resolution, which consequently produces a sharper image.

the pilot with the "big picture." At this point there may be enough information for the pilot to make an informed decision about the general direction in which to deviate to avoid the weather in question.

As the weather is approached and becomes available for analysis on the mid-range scale, the mid-range scale should be selected. Sharper imaging will be displayed and a better interpretation of the weather can be made. Final decisions on the exact flight path route should be made at this time. The pilot should continue to alternate between the mid-range and long-range scale. The majority (i.e., 70% to 80%) of the flight at this stage of the deviation should be conducted using the medium range. However, alternately reselecting the long-range mode will update the big picture and confirm that the action being taken is still appropriate.

Once the flight path deviation route has been determined *and* the weather is within the short-range distance, the short-range mode should be selected. This range provides the highest resolution and the most accurate picture of the weather available and enables the pilot to refine aircraft heading, facilitating the optimum flight path route to bypass the thunderstorm cells. Once again, although the majority of the time (at this stage of the deviation) should be spent operating on the short-range scale, alternating with the mid-range scale will accomplish two things: (1) it provides a smaller version of the big picture, thus reorienting the pilot to the deviation path; (2) it reorients the pilot to the overall weather pattern.

The short-range scale may also be able to depict a hail shaft that might not be shown on a long-range scale. (This will be discussed further in Chapters 2 and 3.)

Although primary attention should be directed to the task at hand, which is guiding the aircraft through the clear areas and alternating between short- and mid-range scales, when possible (time permitting) reselect the long-range mode. Once again this will assist in reestablishing your position relative to the overall weather picture.

Once the area has been circumnavigated, the short-range selection is no longer necessary and the next scale up should be selected as the primary mode of operation. Radar operation should continue, alternating between mid- and long-range scales, until the next weather area is encountered, at which point the entire process is repeated.

Therefore, continuous operation in a single range is neither recommended nor prudent. To obtain the most accurate overall analysis of the weather system, alternating between at least two and sometimes three ranges is suggested.

Essentially, either the mid- or the short-range scale will always be primary — the alternate or secondary scale will always be the next size up.

When weather is anticipated but not yet depicted on the radar screen, the pilot should utilize the *parked* position. The parked position uses the mid-range scale with the antenna tilt adjusted so as to paint ground returns on the top (last) third of the PPI.

There are four advantages in using the parked position:

1. The presence of ground returns serves as a fault light: the absence of ground returns indicates your radar has failed.

2. It ensures that you do not miss any potential weather targets (overscanning).

3. Thunderstorms will automatically be identified. If a target appears in the ground return, "walks out" slightly from the ground return, then disappears, it was either a specific ground feature or insignificant weather.

4. It will identify severe weather by automatically indicating shadowing as the weather is approached.

Let me reiterate one important point. Attempt to finalize your decision on the exact flight path route to be navigated while the weather is still in the mid-range scale — more specifically, when the weather is no closer than 30 to 40 NM. At this point there is enough information for you to make an informed decision. The worst time to formulate your strategy is when you are surrounded by numerous weather cells on the short-range scale. If you are unclear about a safe flight path route while viewing weather on the mid-range scale, don't press on, hoping an answer

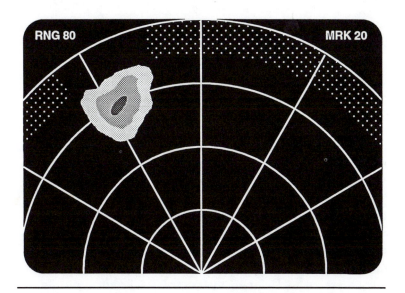

Fig. 1.18 The radar shadow is indicated by the absence of ground returns behind the weather target.

will materialize soon. Exercise conservative judgment and avoid the entire radar signature; make a 180-degree turn if you have to. Don't just let a decision happen — you decide — you *can* make the difference.

Hardware Differences

Although the operation and application of the radar components and controls are essentially the same from unit to unit, some hardware differences merit further exploration. There are three basic types of airborne weather radar:

1. *Analog* has no memory display on the radar screen. It displays weather in various intensities of brightness in one color — referred to as a monochromatic display.
2. *Digital* has a memory function that provides a full-time display with a sweeping update. Weather is displayed with distinct demarcations between shades of one color.

3. *Color* is a digital radar with the ability to display three or more colors representative of various weather intensities.

All radars operate essentially the same way. However, there are some major differences worth mentioning in the hardware used by low-powered and high-powered radars. As discussed earlier, the magnetron significantly boosts the power output in the X band frequency range to 10,000 to 60,000 watts. Unfortunately, the transmissions are not precisely on frequency and tend to drift away from the primary frequency.

Replacing the magnetron with solid-state transmitters solves the frequency irregularity problem but creates another difficulty. Unfortunately, solid-state devices in the X band range are only capable of producing a maximum of 100 to 200 watts of power economically. Therefore, innovative technological improvements are necessary to facilitate the use of solid-state transmitters. These design improvements have come in the shape of high gain receivers and sophisticated signal microprocessors. These sophisticated components used in conjunction with the lower-powered (100- to 200-watt) solid-state transmitters perform as well as the high-powered (10,000- to 60,000-watt) magnetron-based systems. How can this be? A brief review of some of the basic principles discussed earlier will provide some insight.

As you recall, the microwave signal is transmitted at the R/T, distributed by the antenna, reflected back by a target, received by the antenna, processed, and then displayed on the PPI. However, the amount of return signal received is only a tiny fraction of the initial transmitted power pulse. Due to attenuation (see Chapter 2), signal reception is reduced by a fixed percentage at any given distance. Therefore, the quantity of energy received is known and predictable.

Since the energy to be received is a known quantity, any variance in the return signal must be weather (or terrain). The degree of signal return variance is, of course, directly correlated to the weather intensity. Therefore, theoretically, the magnitude of the initial pulse should have no bearing on the efficacy of the radar. Thus, if the signal losses are reduced by a fixed percentage, it really doesn't matter if the transmitter is sending out 60 watts or 60,000 watts.

The radar manufacturer has two basic choices: (1) use a high-wattage/power transmitter with a conventional receiver; or (2) use a low-power transmitter with a sophisticated receiver and a complex microprocessor.

Using design option 2 seems to be the preferred choice for four reasons.

1. Solid-state transmitters have a virtually infinite life span, whereas the magnetron vacuum tubes have a finite life span of approximately 1,000 hours and are quite expensive—about $1,700.

2. Using a frequency with tighter tolerances reduces unwanted, erroneous noise, which accompanies signal return. Additionally, the sophisticated digital signal processors can eliminate random noise. *Noise* is any random energy, whereas a *signal* has a definite pattern. The complex microprocessor statistically identifies and distinguishes the signal from the noise.

3. Using a lower-power transmitter reduces the dangers of arcing and of radar operation around ground personnel.

4. Using a tight frequency transmission facilitates the detection of turbulence via the Doppler phenomenon. Normally, the turbulent area is color-coded magenta on the PPI.

The Doppler effect explains the rise and fall in pitch (frequency) of approaching and departing objects respectively. For example, a listener will observe that as a car approaches, the pitch of the sound rises, and as the car passes, the pitch falls. The frequency shift engendered by the moving object also occurs in other wave phenomena, such as radio waves. Just as the sound of an approaching auto increases in pitch (frequency), the horizontal movement of a raindrop toward an aircraft will also slightly increase the frequency of the energy wave received. The effect is reversed as raindrops move away from the aircraft. Obviously, the horizontal speed of the raindrop is quite small when compared with the velocity of the aircraft. However, in the early eighties, the technology to accurately compute these measurements became available. Generally speaking, a certain horizontal rainfall velocity was determined to be associated with moderate or greater turbulence. When this threshold is reached, the area is displayed on the PPI in the appropriate color.

The effective range of the turbulence mode is limited, and precipitation is necessary for turbulence to be detected. Therefore, use caution and refer to the manufacturer's reference manual for exact details.

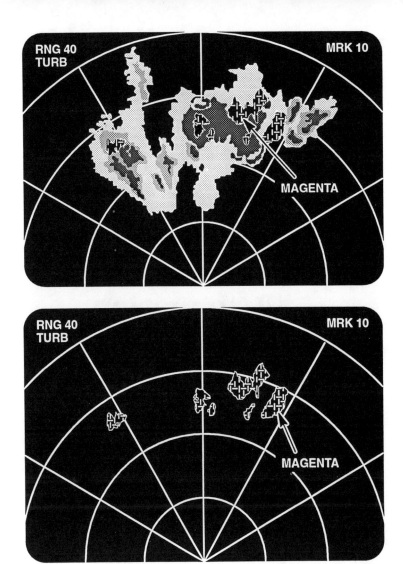

Fig. 1.19 In this figure the turbulence mode has been selected. The upper screen clearly depicts significant weather
and also highlights the areas responding to the Doppler effect
in magenta. The lower screen displays the same target; however, only the areas of magenta are depicted. How can this be
explained? In the lower screen the gain (listening sensitivity)
of the radar is reduced so that weather echoes can no longer
be detected by the R/T. Consequently, the weather display
drops out. The magenta areas remain because the turbulence
function looks at frequency differential, or shift, and not absolute signal strength. Although the return signal is not strong
enough (with the gain reduced) to trigger a weather display
on the PPI, the frequency shift is unaffected. Therefore, the
magenta areas remain intact.

2

Reflectivity and Attenuation

Reflectivity Levels

Now that we have discussed the fundamental theories and principles of radar operation, what exactly will the radar paint (depict or detect) for us? The radar will depict objects that achieve a certain reflectivity level. *Reflectivity* is a target's capacity to reflect microwaves. The higher the reflectivity level, the more pronounced the return will be. As the level of reflectivity diminishes, so will the return. The level is measured in *decibels of reflectivity*. The chart illustrates the reflectivity level of various types of precipitation.

Most Reflective

wet hail

rain

wet snow

dry hail

dry snow

Least Reflective

Generally speaking, the strength of a weather return is directly related to (1) its composition, (2) its size, and (3) the amount of precipitation it contains, with priority given in that order. For example:

Precipitation Form 1		*Precipitation Form 2*	
FORM	dry snow	FORM	rain
SIZE	large	SIZE	medium
AMOUNT	heavy	AMOUNT	moderate

Of these two forms of falling precipitation, the rain is likely to provide a better return due to its higher reflectivity level. In fact, water particles return almost five times the amount of energy as ice particles of similar size. Terrain is highly reflective, with mountainous areas being most reflective.

Other aircraft may be depicted using precise tilt management, but normally this process is time-consuming and has proven to be unreliable in many cases. In addition, the radar is not designed for this type of operation.

The radar will not paint minute cloud droplets, vapor, fog, ice crystals, or small dry hail. Therefore, although you may be flying through a clear area on the radar, you may in fact be in instrument meteorological conditions (IMC). This is a classic scenario when operating in the vicinity of *embedded* thunderstorms. In this

Table 2.1. Thunderstorm intensity and rainfall rate

VIP Level	Echo Intensity	Precipitation Intensity	Rainfall Rate (in/hr) Stratiform	Rainfall Rate (in/hr) Convective
1	weak	light	less than 0.1	less than 0.2
2	moderate	moderate	0.1–0.5	0.2–1.1
3	strong	heavy	0.5–1.0	1.1–2.2
4	very strong	very heavy	1.0–2.0	2.2–4.5
5	intense	intense	2.0–5.0	4.5–7.1
6	extreme	extreme	more than 5.0	more than 7.1

SOURCE: FAA Advisory Circular 00-45c.

Echo intensities range from level 1 (weak) to level 6 (extreme) storms. These readings are predicated on ground-based meteorological weather radar. The information is then obtained from the video integrator processor (VIP) and is expressed as a VIP level (e.g., VIP level 3). Echo intensities assigned as VIP level 3 or higher should be considered extremely hazardous. A level 3 storm is classified as strong and will depict contour (red) on the radar screen.

case, you will lose visual contact with the storm—all the more reason to have a solid understanding of airborne weather radar operation and interpretation.

An inference can be made from the preceding material. The raindrop size is almost directly proportional to the rainfall rate. As these two variables increase, reflectivity will increase. Rainfall rate is directly related to thunderstorm intensity (storm level). It can be inferred then that the areas of highest reflectivity are representative of the strongest storms.

Turbulence probability can be directly correlated with the reflectivity level. Level 1 storms pose a 10% chance of moderate turbulence. Level 2 storms pose a 40% chance of moderate turbulence with a 3% to 4% chance of severe turbulence. Once a storm reaches level 3, there is a 50% or better risk of severe turbulence. In addition, when the storm matures to this level, the corresponding turbulence is not only associated with the areas of maximum reflectivity but may be present in any part of the radar *signature* (i.e., associated with any part of the storm system).

Destructive hail is also directly associated with the storm level. Consider damaging hail present in all level 3 storms or higher. As with turbulence, when a storm reaches level 3, hail may be located in any area associated with the storm. Hail shafts may also extend out, away from the storm. The shaft itself can take on any dimension but in this context is characterized by a localized stream of hail. Even if the shaft is composed of wet hail (which is highest in reflective capability), it may still go undetected on the radar. There are some thunderstorm shapes, however, that are indicative of hail (see Fig. 2.1).

Although hail shafts are difficult to detect, on occasion (through proper tilt management and using the short-range scale) a shaft may appear as a small dot on the radar in the proximity of a level 3 or higher storm. Techniques to detect these shafts will be discussed in Chapter 3.

Fig. 2.1 (*a–d*) These unusual shapes have been associated with the presence of hail. **a**, finger; **b**, hook; **c** scalloped edge; **d**, U-shape.

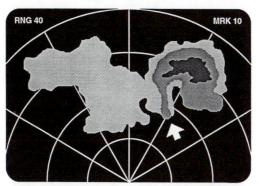

a

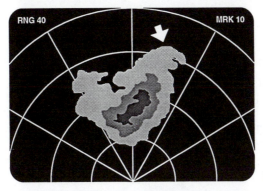

b

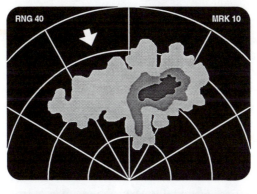

c

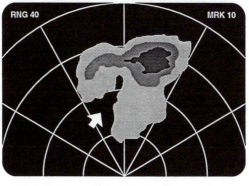

d

Thus, we have established an easy-to-remember correlation regarding reflectivity:

Reflectivity is correlated to raindrop size
which is correlated to rainfall rate
which is correlated to storm level
which is correlated to turbulence and hail.

Attenuation

Attenuation is the loss of signal strength or signal return. Attenuation occurs for several reasons, and we will discuss each of them. It is absolutely imperative that the pilot be able to correctly identify the telltale signs of attenuation. Failure to do so may have life-threatening ramifications.

Transmission Losses

Even on a clear day, as soon as the microwave energy leaves the antenna, the signal begins to attenuate. In fact, signal attenuation is proportional to the square of the distance the signal travels. For example, if you are depicting a storm at 15 NM, an identical storm at 30 NM will be displayed at ¼ the intensity of the cell at 15 NM, a cell at 45 NM will be displayed at 1/9 the intensity, and so forth (see Fig. 2.2).

To compensate for this type of attenuation, engineers have developed a *sensitivity timing control* (STC), which is a type of volume control on your listening device (receiver). As the distance of a target increases, the STC increases the listening capability of the receiver. So in our case, the receiver is listening four times as hard to the target at 30 NM as it is to the one at 15 NM. Because of the STC, identical targets in this scenario will be displayed at identical intensities.

However, there is a major limitation to the STC. Unfortunately, the volume (or sensitivity of the receiver) cannot be turned up indefinitely. In fact, the volume maximum is normally at a range of 30 to 40 NM. This distance is defined as the STC range.

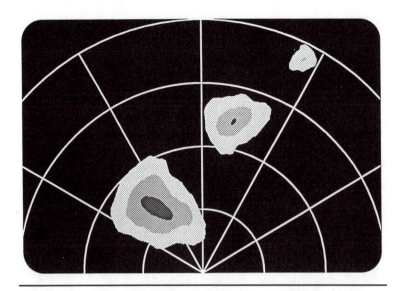

Fig. 2.2 Three storms of equal reflectivity are displayed in various intensities due to signal attenuation over distance.

Thus, echoes within the STC range will be relatively representative of the true storm intensity. But more important, storms outside this range will be depicted *smaller* and *weaker* than they really are.

Shadowing

The second type of attenuation concerns the forces at work when the energy strikes a target. When a microwave pulse strikes a target, a portion of the energy is reflected back. Attenuation occurs when a portion is absorbed by the target itself or a portion passes through the target, perhaps to reflect off of another target. Significant target density (as found in heavy weather) will tend to absorb the majority of the signal, leaving very little for return and none for penetration. This type of attenuation is referred to as *shadowing*.

Varying degrees of shadowing can occur. The most common is normally associated with some type of contouring cell that appears to be accurately displayed. However, even though the returning signal may be providing a fair description of the storm, too much of the signal is being absorbed. No energy is left to detect weather behind the storm or produce ground returns. To confirm this, tilt the antenna down so that ground returns are displayed at approximately the same distance as the target. A shadow (i.e., no ground returns) appearing behind the heavy weather tells you that the storm is blocking your radar's view of what lies behind the storm.

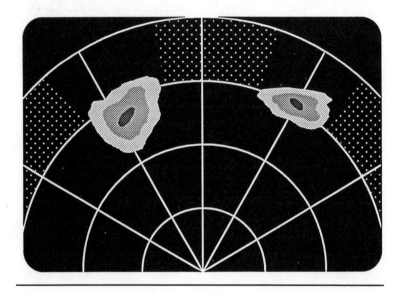

Fig. 2.3 Both of these cells are creating shadows. The intensity of these targets is such that no ground returns are produced behind these thunderstorms.

Another variation of radar shadowing occurs when the storm is *so* heavy it now absorbs practically all of the signal. Little energy is left and a fair display of the storm is unobtainable. The classic characteristics of this type of attenuation are a crescent-shaped target, concave on the back side, and no ground returns beyond the target. Do not misinterpret this area for a thin spot in the storm. It is, in fact, the area of heaviest weather.

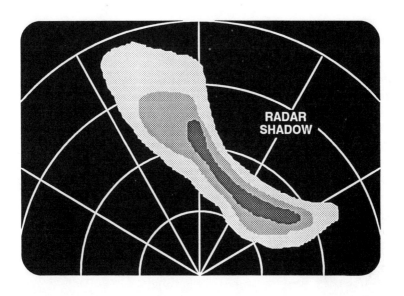

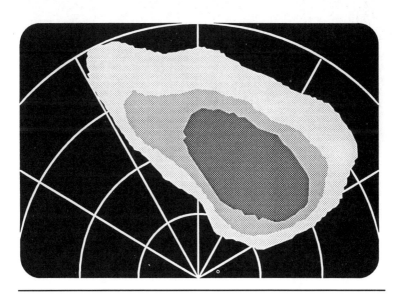

Fig. 2.4 The radar screen on top shows how an attenuated signal may be portrayed. If the radar could detect the true intensity of the storm, the return on the bottom screen would be displayed.

Perhaps the most insidious form of radar shadowing occurs while in a widespread area of moderate or heavy rain. Typically, the rain appears to stop at the 20-NM mark or less. However, as the flight progresses, the end of the area of rain gets no closer. The distance between the aircraft and the outer perimeter appears fixed. This condition is particularly dangerous when *embedded* thunderstorms are present. The intervening weather is attenuating the signal to such an extent that even though the rain may extend for 80 NM or more, only the first 20 NM or so will be depicted. Obviously, severe weather may lie beyond the detectable range. To resolve this nebulous display, tilt the antenna down in an attempt to produce ground returns—and fly toward them. The areas that produce ground returns mark the areas of least signal attenuation, that is, areas of lesser rain. If ground returns cannot be painted, advise Air Traffic Control (ATC) that your

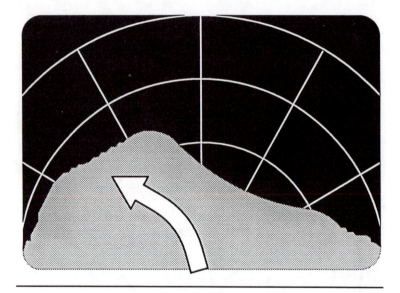

Fig. 2.5 In this scenario a deviation to the right would appear to be the most expeditious course out of the weather—when in fact it may be leading directly into an area of heavy weather. Note the similarity in the concavity of this figure and the top drawing in Fig. 2.4.

radar capabilities are limited, and obtain vectors away from areas of weather.

Although you should rely mainly on your weather radar and not ATC's, options under these circumstances are limited. In addition, if you observe a bulge (toward you) at the outer perimeter of the rain, *do not fly toward it*. Fly instead to the areas generating the farthest returns. In this case a cell is positioned at the bulge, absorbing more signal (producing increased attenuation) and providing less return.

Radome Icing

Radome icing can also cause attenuation. The radome is a honeycombed structure covering the antenna, which is normally located at the nose of the aircraft. It is tuned to the radar frequency and should have a high transmissivity level. Astro coat is a special paint applied to the radome so transmissivity is not decreased. A coat of ice or rain on the radome can reduce system performance. Severe radome icing can cause targets to drop out completely. This may be accompanied by a haziness on the PPI emanating from the zero distance marker outward.

Reception Losses While in Weather

The last form of attenuation occurs while painting weather *in* weather. Remember, airborne weather radar was designed for weather *avoidance*, not weather penetration. Any target displayed while in weather will be *understated*.

Summary

The salient points in this section are:

1. Be cognizant that any weather depicted beyond the STC range (approximately 40 NM) will be understated.
2. Weather painted while *in* weather will be understated.

Airborne weather radar is for weather avoidance, *not* weather penetration.

3. Never, *ever* fly toward a radar shadow. Never plan to deviate around a target creating a radar shadow and then use that shadow area as your only means of escape; you do not know what lies there.

4. If, for some reason, attenuation is pervasive (as is the case in a widespread area of moderate or heavy precipitation), *fly toward ground returns* or the farthest return. Do not fly toward areas that bulge toward you. Think of the bulge as "coming to get you."

When a decision is unclear, always fly toward areas of ground return. An easy way to remember this is to just think that at this point of the flight you are probably wishing you were on the ground; therefore, follow the ground returns.

3

Tilt Management

Antenna Beam

The single most important key to accurately analyzing and synthesizing the information on your radar is proper *tilt management*. Tilt management is probing weather with the microwave beam to obtain an accurate assessment of the weather. Notice I use the word *proper* and not *precise* when referring to tilt management. Even good tilt management cannot provide precise manipulation of the radar beam. Due to the vagaries in the system, do not expect to do optic laser surgery with your radar. *Do* expect to direct the beam within a couple thousand feet of the desired position. This will enable you to obtain an accurate assessment of the weather conditions. However, before you can manage the beam, its characteristics must be clearly defined and internalized.

The radar beam is microwave energy in a cone-shaped (pencil) beam, similar to that projected by a flashlight. The beam itself never changes. Pursuing an analogy at this point will prove beneficial. If you were to stand at the end of a dark room and shine a flashlight on the opposite wall at a picture, for example, not only would the picture be illuminated but so would the entire wall. The width of the beam at the wall is quite wide. Now, maintain your position and ask someone to step 3 feet in front of you. The width of the flashlight beam at that point is probably only as wide as your assistant. Has the beam itself changed? *No.* Has the width of the beam changed? *No.* The beam was always that wide at the distance your partner is standing. Therefore, the width of the beam is constant at any given point, but each point along the beam has a different beam width. If you fully understand the last sentence, you have the idea. Thus, distance (or range) along the beam influences beam width.

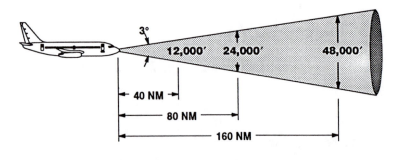

Fig. 3.1 This diagram shows that the beam width and distance are directly related. Beam width is constant at any given point, but each point along the beam has a different beam width.

The other factor influencing beam width is antenna size, which is responsible for the angle of the beam. Antenna size and beam angle are inversely related. In order to calculate the beam width, you must know the angle of the beam and the distance along the beam.

The formula for calculating beam width in feet at any given distance is

number of degrees in antenna beam × distance (NM) × 100 = beam width dimension (feet)

For example, given an antenna beam angle of 3 degrees, the beam width at 50 NM is 15,000 feet:

3 degrees × 50 (NM) × 100 = 15,000 feet

Another method of determining beam width (one that the author prefers) is calculating the beam on a degree per degree basis. The formula is

Distance (NM) × 100 = beam width in feet per degree of beam angle

Therefore,

50 NM × 100 = a beam width 5,000 feet wide per degree

If you have a 3-degree antenna, then the beam width is 15,000 feet (3 degrees × 5,000 feet) at 50 NM.

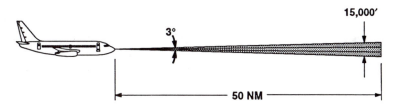

Fig. 3.2 The width of a 1-degree beam at 50 NM equals 5,000 feet. The width of a 3-degree beam at 50 NM equals 15,000 feet (3 × 5,000).

This formula is actually an extrapolation of one that is more common and usually introduced early in pilot training. A standard rule of thumb, when navigating with very high frequency omni ranges (VORs), is: If the aircraft is 1 degree off course at 60 NM, the aircraft is 1 NM (6,000 feet) off course. The above formulas can be applied.

60 NM × 100 = 6,000 feet per degree of beam angle

In the example below, the beam angle is 1 degree; therefore,

6,000 feet × 1 degree = 6,000 feet, or approximately 1 NM

Therefore, at 30 NM the beam width of a 1-degree beam is 3,000 feet. Refer back to the radar antenna beam shown in Fig-

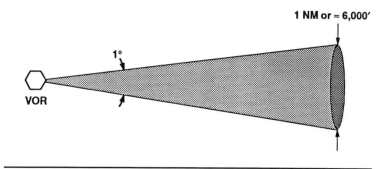

Fig. 3.3 The calculation of VOR beam width is just a variation on the method used in calculating the beam width of our weather radar beam.

ure 3.1. It depicts the beam width at various points along the beam. Calculating beam width is one of the prerequisites for understanding tilt management.

Beam Filling

The radar is calibrated to measure a certain percentage of reflectivity. In order for the radar to depict a target on the PPI, a certain amount of the return must be received. If only a low percentage of the microwave is picked up, the target reflecting that energy will appear on the PPI as a weak echo, *regardless of its actual strength*. Conversely, if a high percentage of the return is received, the target will appear strong and well defined, thus providing a more accurate assessment of the true nature of the target. A weather area's capacity to fill the *entire* width of the beam is directly related to the percentage of reflectivity received. Essentially, if the beam is wider than the target being painted — that is, if the beam is only partially filled by the target — the weather displayed will be *understated*. The storms are stronger than they appear.

The farther away a target is, the less likely it will be able to fill the beam. For instance, the width of a 3-degree beam at 150 NM is 45,000 feet. Therefore, a 35,000-foot thunderstorm would

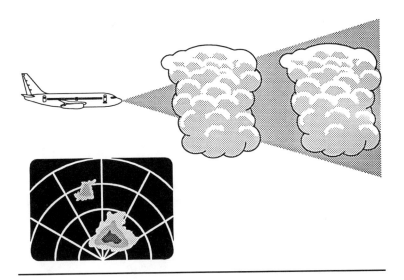

Fig. 3.4 These two thunderstorms are identical. However, the more distant target does not fill the beam and is therefore understated on the PPI.

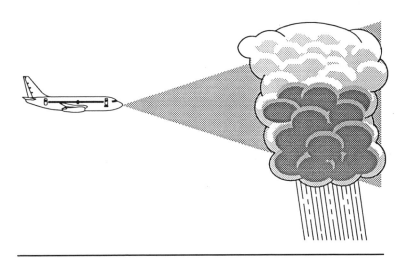

Fig. 3.5 This target fills the entire beam. Unfortunately, it is quite possible that the reflective core of the storm will not fill the entire beam and therefore the return will be understated.

not fill the beam and consequently would be understated on the PPI. Furthermore, the upper 20% or more of a TRW may be composed of nonreflective elements such as dry hail. Therefore, the TRW may actually have a vertical profile of 40,000 to 50,000 feet with only the lower two-thirds (or 35,000 feet) responsible for reflective moisture.

The most important point here is that whenever a target is depicted at approximately the 80 to 100 NM range, it will typically be understated because the entire storm (as a rule) cannot fill the beam. The target will be further understated because of the attenuation that occurs at this distance.

However, even if a cell is much closer (e.g., 40 NM), improper tilt management may still prevent the storm from filling the beam and therefore it will be understated. This can lead to *misinterpretation of dangerous weather.*

Proper tilt management is the single most important skill a pilot can develop to obtain more accurate displays on the airborne weather radar. Figure 3.6 shows a form of *overscanning*. (*Under-*

Fig. 3.6 This target is close enough to the aircraft for it to easily fill the radar beam. However, improper tilt management (in this case directing the beam well above the most reflective area of the storm, i.e., overscanning) is grossly understating the target.

scanning is equally dangerous and normally occurs when the aircraft is on the ground or in the low-altitude structure.) Although the storm is much wider than the beam at this distance (see Fig. 3.6), the aviator has selected an antenna position that merely clips the top of the storm. This results in only a small percentage of reflectivity, and consequently, this severe cell is displayed as only light precipitation. A few small adjustments of the antenna tilt would enable the pilot to recognize the true potential hazard of this storm.

Proper tilt management requires knowing the beam width at any given range. For example, in Figure 3.6 the target is 40 NM in front of the aircraft (which is utilizing a 3-degree beam). The beam width is 12,000 feet at 40 NM (40 NM × 100 × 3 degrees). Now that the beam width has been calculated, we must establish the *position* of the antenna. In our flashlight analogy, the beam was always clearly visible with the naked eye. Thus we were aware of its position. Microwave, however, is invisible. How, then does the pilot know if he is painting the lower, middle, or upper section of the storm? This information can be obtained by establishing a *ground base*.

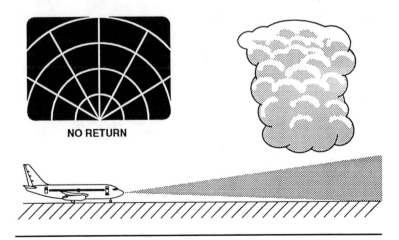

NO RETURN

Fig. 3.7 The aircraft on the ground is underscanning the target directly ahead. Consequently, no return is displayed on the PPI. Obviously, this weather misinterpretation is extremely dangerous.

If a flashlight beam is directed at the floor and the distance is measured from the flashlight to the area where the beam strikes the floor, then, as the beam is raised 1 degree at a time, its distance relative to the floor can be calculated. If the pilot tilts the antenna down, symmetric ground returns should begin to appear at the far end of the PPI. (This is analogous to the flashlight beam sweeping the floor.) As the tilt continues down, the ground returns move closer to the aircraft. In our example, the target is at 40 NM, so move the tilt down so that ground returns appear at the 40-NM range arc. We now know the position of our beam. We also know that 1 degree at 40 NM is equal to 4,000 feet. Therefore, if we move the antenna tilt knob (which is calibrated in degrees) 1 degree upward, we know we have moved or lifted the bottom of the beam up 4,000 feet. (Note: Ground returns have moved farther away.) Thus, we know the bottom of the beam is penetrating the storm at 4,000 feet above ground level (AGL), and the top of the beam ends at 16,000 feet AGL (4,000 feet +

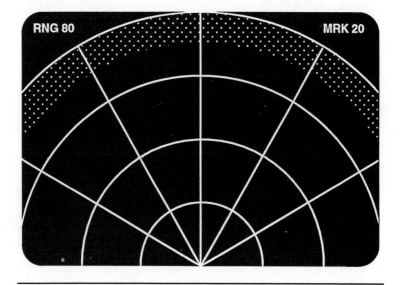

Fig. 3.8 As the antenna is tilted down, ground returns begin to appear (in this case at approximately 70 NM). This is the first step in determining antenna beam position.

beam width of 12,000 feet). Tilt the antenna up another degree and the bottom of the beam is moved up (at 40 NM only) another 4,000 feet. Now the bottom of the beam is striking the cell at 8,000 feet AGL and the top at 20,000 feet AGL.

Using this technique, a very rough approximation can be obtained of the TRW tops. In the previous example, we were tracking the beam as it was lifted off the ground 1 degree at a time. This 1 degree was then converted to feet. In addition, as the beam was raised, the depicted weather changed in both intensity and shape. Most likely, as the beam was raised, the echo intensity increased and the shape of the thunderstorm was well defined (identifying the storm's core reflectivity area). However, as the beam is lifted further, the return begins to fragment and eventually fades away, leading the pilot to believe he has reached the top of the storm. It may be *tempting* to assume that the entire height of the storm can be determined by calculating the number of degrees the beam was raised from the ground to the point where returns dissipate. But remember, severe turbulence and dry hail may lie above the detectable area of the storm. The true height of the storm may be as much as one-third or more greater than the height calculated using your airborne weather radar. For convenience, let us call these calculated but inaccurate thunderstorm tops *radar tops*.

A by-product of computing radar tops is the ability to determine weaker areas of weather. As you tilt the antenna up, the weaker storms will normally be the first to drop out of the PPI. Reducing the gain will confirm these weaker areas of weather.

If determining the tops of thunderstorms with your airborne weather radar provides such an unreliable approximation of thunderstorm tops, then why use it? There are two reasons. First, it does provide at least a rough approximation of the thunderstorm height. As a general rule, if your altitude is two to three times the calculated radar tops, you should be able to overfly the area safely. Second, and most important, computing the radar tops incorporates all aspects of antenna tilt management. During this process, the pilot is forced to do the following:

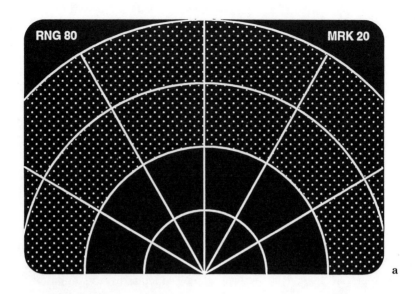

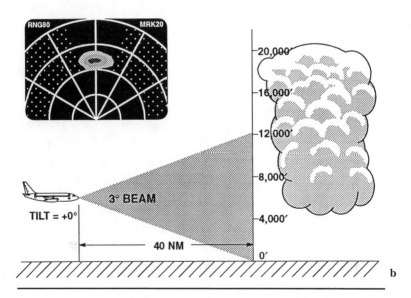

Fig. 3.9 (*a–d*) As the process progresses, antenna tilt continues down, and ground returns move closer to the aircraft (3.9**a**). This establishes a *ground base*. Figures 3.9**b**, 3.9**c**, and

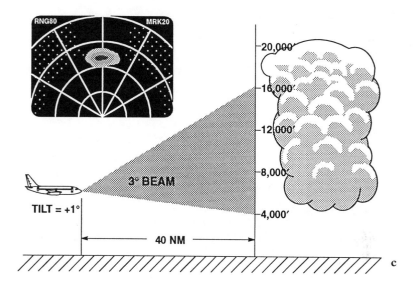

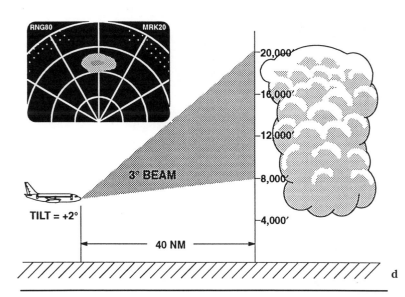

3.9**d** graphically illustrate antenna beam movement. In this case, each 1 degree of tilt vertically displaces the beam 4,000 feet (at 40 NM).

1. Depict ground returns (see section on range in Chapter 1).
2. Determine the beam width.
3. Determine the beam position.
4. Determine if the target is beam filling.
5. Slowly probe the target (evaluating shapes, gradients, etc.).
6. Identify radar shadows (indicative of heavy weather).
7. Identify weaker areas of storms.

Beam Width Smearing (Azimuth Resolution)

Whenever the distance between two targets is less than the radar beam width at that distance, weather will appear worse than it actually is. This is the *only* time weather severity will be overstated.

As the cone-shaped energy is transmitted out and sweeps the area, weather will first be detected by the edge of the cone. At that point the radar "thinks" weather is at the middle of the beam and it will be depicted as such. Thus, the targets will be "smeared" on the PPI as the leading edge of the cone detects one target early and the trailing edge of the cone detects the other target late.

As illustrated in Figure 3.10, even though there are actually two distinct cells, the radar merges them, producing one large image on the PPI. *Beam width smearing*, then, is a function of the antenna's beam width and the range of the target. As the weather cells are approached, they appear to separate and, consequently, take on their true shape. Beam width smearing is more likely to occur when targets are farther away (since the beam is wider, thus offering a greater opportunity for two targets to fall within this distance). We already know that targets painted at distant ranges are *understated* and, therefore, should be completely avoided. Beam width smearing does not provide practical information for your course deviations, because you may not know if smearing is occurring until you are too close to do much about it.

Don't assume beam width smearing when weather is being

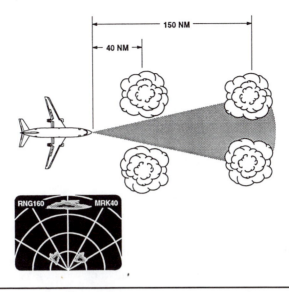

Fig. 3.10 Beam width smearing (azimuth resolution) occurs whenever the distance between two targets is less than
the radar beam width at that distance. The lateral separation
between the two more distant targets is less than the beam
width at that distance. Azimuth resolution occurs, and the
targets at 150 NM appear as one large target rather than two
separate cells.

depicted at 50 NM or less. At this point, your beam is so narrow
that minimal smearing can take place.

Beam width smearing is an anomaly in the radar system, but
don't misconstrue this phenomenon as a radar malfunction.

Zero Tilt

Calibrating True Zero Tilt

A final formula to add to our tilt management program deter-
mines the *zero tilt* position on the radar system. Zero tilt positions
the antenna so the center of the beam is projected parallel to the
surface of the earth.

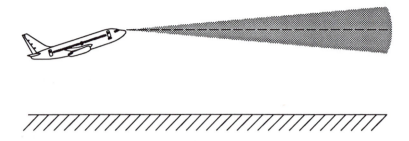

Fig. 3.11 If the tilt control is calibrated accurately, setting the antenna to zero tilt should position the center of the beam parallel to the surface of the earth.

Although all radar antenna tilt knobs have a zero tilt mark, many may be miscalibrated. Therefore, just because you have selected zero tilt, do not be misled into thinking that your antenna is at the zero tilt position. Determination of *true zero tilt* can be calculated by using ground returns and the zero tilt formula.

The zero tilt formula is:

$$\frac{(\text{altitude}/100)}{(\frac{1}{2}) \times (\text{beam angle})} = \text{the distance (in NM) that ground returns should begin at the zero tilt setting}$$

The formula looks more complicated than it is; let us simplify it.

1. Take two zeros (00) off your altitude.
2. Divide that number by half your beam angle.

Therefore, if you were cruising at 20,000 feet and were utilizing a 4-degree beam, the true zero tilt position should begin to paint ground returns at approximately 100 NM.

$$\frac{(20,000/100)}{(\frac{1}{2}) \times (4)} = \frac{200}{2} = 100$$

Another example: You are cruising at 10,000 feet using a 3-degree beam:

$$\frac{(10,000/100)}{(½) \times (3)} = \frac{100}{1.5} = 66 \text{ NM, or approx. } 50 \text{ NM}$$

Remember, this is not an exact science; so if your antenna beam is an odd number, round the denominator (1.5 in this case) up to the next whole number. If you want to be more precise (and you use the same antenna or aircraft all the time), then compute a figure once and always check for ground returns at that altitude.

If the zero tilt mark on your radar does not produce ground returns (in the last case) at approximately 66 NM, then position the tilt so as to achieve ground returns beginning at about 66 NM. This is the real or true zero tilt setting on the radar system, and a mental note should be made identifying this position. If returns are present where they should be, within plus or minus 1 degree of the zero tilt mark, consider your radar to be calibrated within acceptable limits. If this is not the case, then make the appropriate adjustment/correction mentally and position the antenna there whenever zero tilt is desired.

The zero tilt calculation is most accurate when determined at a lower altitude. For instance, in our above examples, the zero tilt setting determined at the lower altitude would be more accurate than the one calibrated at the higher altitude. Ground returns at 66 NM will be more pronounced and defined than ground returns at 100 NM. Remember, as distance increases, the beam width increases. The more a target (i.e., the ground) fills the beam and the more reflective it is (and the ground is very reflective at shorter ranges), the more pronounced it will be. The capacity of the ground/terrain to fill the beam is reduced as distance increases. The *only* way to increase the amount of ground filled by the beam is to tilt the antenna downward.

In fact, at greater distances, significant ground targets have more difficulty filling the beam than significant weather at the same distance. This is because thunderstorms are *vertical* targets, enabling the beam to strike at close to the perpendicular,

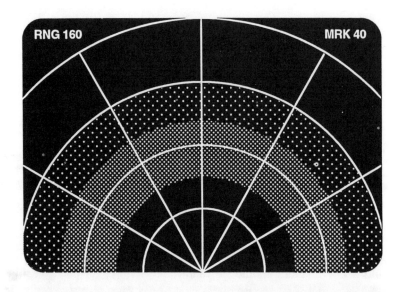

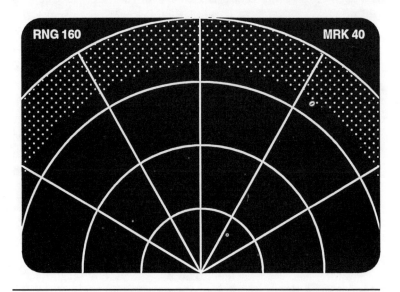

Fig. 3.12 The screen on the top has pronounced ground returns beginning at 60 NM. The screen on the bottom has ground returns of lesser intensity due to the difficulty in filling the beam at greater distances.

whereas ground targets are *horizontal* targets struck by the beam at glancing angles close to the parallel. The angle between the beam and the target is referred to as the *angle of incidence*. A smaller incident angle (target with a greater vertical profile) will engender a more pronounced return than a target with a larger incident angle (target with a more horizontal profile). The terminology, in this case, is not nearly as important as the concept. Therefore, it may prove beneficial to take a few extra minutes to review the illustrations carefully to internalize this whole idea.

Substantive ground returns beyond 80 to 90 NM are difficult to obtain. In some cases, a portion of the ground returns can be detected as far out as 200 NM. However, at this distance the return is degraded not only because of a large angle of incidence but also because of the curvature of the earth. Beyond 200 NM the curvature of the earth actually departs the bottom of the beam.

Use of Zero Tilt on the Ground

Operating the antenna at the true zero tilt position in the high-altitude structure does not offer any significant advantages. Using your radar tops calculation provides you with ample infor-

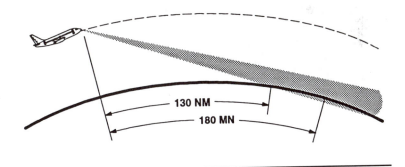

130 NM

180 MN

Fig. 3.13 This figure illustrates three important points when attempting to paint ground returns at great distances. (1) The beam width is too large to be filled. (2) The incident angle is quite large, resulting in a poor return. (3) Beyond approximately 200 NM the curvature of the earth actually departs the bottom of the beam.

mation. However, zero tilt is very useful in weather interpretation while the aircraft is on the ground or in the low-altitude structure.

We would use our zero tilt knowledge while taxiing out to enhance our assessment of the weather in the departure area. If the antenna is calibrated properly, positioning the tilt to 0 degrees on taxi out, or just prior to departure, provides a defined baseline above which the antenna should be directed. In order to detect weather in the immediate vicinity of the aircraft (while on the ground or in the low-altitude structure), approximately 5 to 10 degrees of up-tilt are required. A tilt setting of 0 degrees would tend to underscan the target; this could lead to dangerous misinterpretations of the weather.

Even though the beam can get quite wide, it will not be wide enough to detect hazardous weather in the immediate vicinity of the airport unless the beam is directed upward (above zero tilt). The weather of primary concern will most likely be that weather closest to the airport. The beam width in the vicinity of the airport is quite narrow and, therefore, less likely to detect targets. Additionally, if the antenna tilt is set at zero, only the upper half of the beam will be available to detect weather. In order to detect

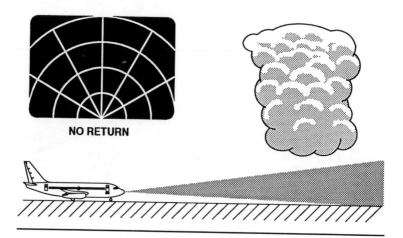

Fig. 3.14 The aircraft on the ground is underscanning the target directly ahead. Consequently, no return is displayed on the PPI. Obviously, this weather misinterpretation is extremely dangerous.

Fig. 3.15 The aircraft on the ground has the antenna set at 0 degrees tilt. Only the upper half of the beam, if that, is available to detect weather. Due to poor antenna tilt management, a significant cell directly ahead appears understated on the PPI.

the weather, the beam must be raised to scan for targets that are actually above the aircraft (as compared with targets in cruise whose reflective core is primarily below the aircraft).

Tilt the antenna up until the thunderstorms are displayed. The shorter range should be selected. As weather is detected, probe the target, just as we did in determining radar tops. Select contour and continue probing the weather, analyzing gradient, shape (discussed later), and approximate height. Incorporating our radar tops technique, we can calculate a rough approximation of the size and height of the storm to assist in determining its overall intensity. This approximation can be calculated by using the main axis (center) of the beam as a point of reference. If zero tilt is calibrated accurately and the zero tilt setting is selected, then the full length of the main axis is essentially the same distance above the ground as the aircraft. In the case above, the aircraft is taxiing for takeoff; therefore, the main axis of the beam is at ground level. If a target lies at 30 NM and positioning the antenna at 7 degrees of up-tilt generates significant returns on the PPI, this weather is clearly hazardous. Once again, the width

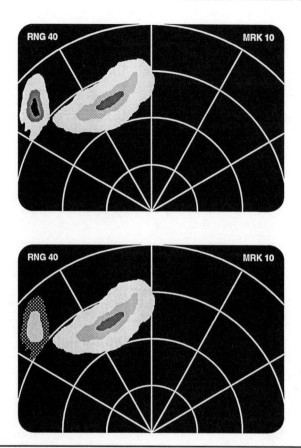

Fig. 3.16 As the gain is reduced, the weakest returns drop out.

of 1 degree can be calculated by taking the distance in nautical miles, multiplying by 100, and converting that number to feet. In this case, each degree at 30 NM equals 3,000 feet. Since we have raised the beam 7 degrees, the center of the beam (at 30 NM only) is 21,000 feet above the ground (3,000 feet × 7 = 21,000 feet). Therefore, the top half of the beam is above 21,000 feet and the bottom half of the beam is below 21,000 feet. This is not done for the purpose of attempting to top the weather. Your aircraft does not have the takeoff performance to top any weather this close to the field.

Remember, steep gradients are considered more hazardous than shallow gradients. As the area of weather is probed with the antenna, both the shape and gradient will change. If a number of thunderstorms are present, identifying the weakest area may be most useful. Our gain switch can be used to determine relative intensity. As the gain is decreased (remember, the gain controls the "listening sensitivity" of the receiver; this sensitivity is decreased when the knob is rotated counterclockwise), the weakest returns will begin to drop out, fragment, and become less well defined. The stronger areas will tend to maintain their shape and definition.

After you have analyzed the weather in the immediate vicinity, select the next longest range for long-term planning considerations. Doing this may greatly influence your departure path strategy. For example, on viewing the return shown on the short-range scale in Figure 3.17, a deviation to the right may appear to be appropriate. However, after the weather is viewed on the longer-range scale, it can be seen that deviating right may actually create more problems. As illustrated in Figure 3.17, if you deviated to the east and your actual route was to the west, you might encounter extensive delays while airborne trying to rejoin your route. Many factors will influence your final decision regarding a scenario like this. (These will be covered in Chapter 5.)

In this scenario, we have assumed the zero tilt position is properly calibrated. If we fly the same ship number all the time (i.e., the same aircraft, not just the same type), we will know if the antenna tilt is properly set. If the tilt is miscalibrated, we will be aware of the correction to make on the tilt knob in order to obtain zero tilt. However, if you must change aircraft frequently, this information may not be known. Unfortunately, we cannot use our zero tilt formula because it requires us to have altitude and depict significant ground returns, neither of which is available on taxi out. Therefore, the full range of the antenna tilt may be required to depict the weather in question. Be very circumspect of radar units that require you to use tilt-down input before producing weather returns while you are still on the ground. In this case, your antenna is grossly miscalibrated. In addition, be cautious of units that require 10 to 15 degrees tilt-up before display-

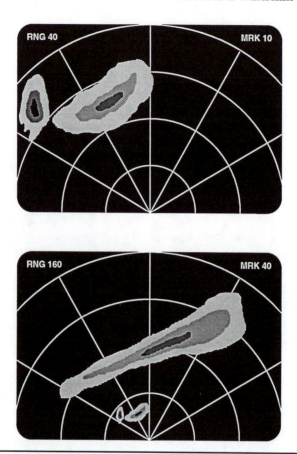

Fig. 3.17   There is no substitute for long-range planning. Selecting the long-range scale provides the pilot with the big picture. Consideration of other options is in order and decisions can be made accordingly.

ing any weather. The only time this setting should be used is to detect very *high base clouds* that contain rain showers or thunderstorms.

Use of Zero Tilt on Approach

Knowing the zero tilt position is also quite useful while on approach. During this phase of flight, the most reflective area of

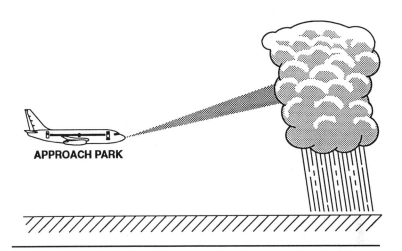

Fig. 3.18 During low-altitude or approach operations the most reflective part of the storm may well be above the aircraft. In this case antenna up-tilt would be required in order to depict the storm.

the weather cell (or cells) may, in fact, be above the aircraft. In general, most altitudes prescribed on the approach chart will be 1,500 to 5,000 feet AGL.

As illustrated in Figure 3.18, the proximity of the aircraft to a vertically built cell would be depicted on the PPI only if the antenna was tilted up toward the reflective area of the storm. If this is not done, the chances of underscanning a target are significant. This is especially true of a target that is close (0 to 30 NM) to the aircraft. At this distance, the beam is so narrow that a target may easily go undetected if proper tilt management is *not* exercised.

Summary of Zero Tilt and Radar Tops Techniques

We can now begin to formulate a general strategy that incorporates the zero tilt and radar tops techniques. The format of this approach divides the vertical airspace into *low-*, *mid-*, and *high-*altitude structures. Generally, if you are operating in the low-

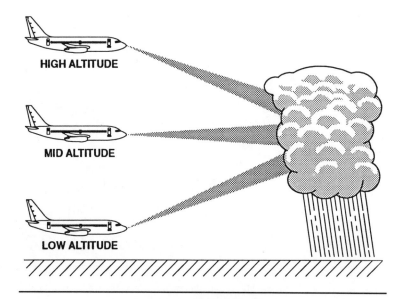

Fig. 3.19 This figure clearly illustrates the inverse relationship between antenna tilt and aircraft altitude. In order for the three aircraft to acquire a similar account of the same storm, three different tilt settings are necessary.

altitude structure, defined as the ground to 5,000 feet AGL, knowing the zero tilt position and scanning upward are critical. In the mid-altitude structure, 5,000 to 15,000 feet AGL, combine both the zero tilt, with corresponding tilt-up increments, and the radar tops technique. Attempt to paint ground returns when able. When in the high-altitude structure, above 15,000 feet AGL, the radar tops technique will offer the most information. The inference here is that aircraft altitude and antenna tilt are inversely related.

We are now beginning to incorporate a number of tools the radar offers to determine the severity of a storm. We know how to use the range control to plan long-term strategies, deviate around close-in weather, and recognize severe weather. The contour function identifies rainfall intensities. The gain switch can assist in determining the relative intensity of the storms depicted, thus identifying the more severe areas. Finally, antenna tilt management is the real key to accurate weather evaluations and, consequently, is tied in to everything we have discussed. If you do not understand the section on tilt management (i.e., antenna beam shape, beam filling, radar tops, ground returns, zero tilt, etc.), your radar set will have little value for you. Proper tilt management is essential. The tilt control should be the most frequently used control on the radar set. As your tilt prowess develops, some interesting returns may begin to take shape.

Target Shapes

Shapes of weather returns depicted on the PPI may be indicative of certain weather phenomena. Some shapes to watch for are

1. rapidly changing echo or echoes,
2. pendant-shaped storms,
3. storms with scalloped edges,
4. V- or U-shaped cells,
5. "hooks" within cells,
6. storms with "fingers,"
7. storms with asymmetric shapes,
8. storms with thin, curling shapes constantly changing,
9. V shapes that appear to be just a layer of rain,
10. a small dot displaced from, but in proximity to, a thunderstorm,
11. building or broken lines of thunderstorms, and
12. angular targets.

Although we may rely on the contour feature and significant attenuation (i.e., shadowing) to identify hazardous weather, the presence of the above shapes should be equally alarming. Let us review some of the particulars of these shapes.

1. Rapidly changing echoes can be indicative of very unstable air with significant vertical gusts and/or the formation of a severe thunderstorm.
2. Pendant-shaped storms have a blunt end and a pointy

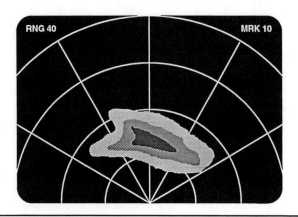

Fig. 3.20 Pendant-shaped storm.

end. They tend to be typical of thunderstorms that produce tornadoes. This is especially true with thunderstorm lines oriented from the northeast to the southwest.

3. Scalloped edges may be indicative of vertical gusts, hail, and turbulence, which can all be present even though the PPI displays no heavy returns.

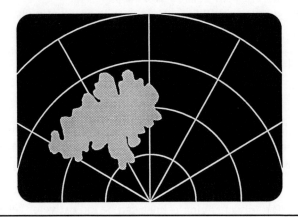

Fig. 3.21 Scalloped edges: Although the PPI displays no heavy returns, vertical gusts, hail, and turbulence may be present.

4. V and U shapes are associated with columns of strong vertical movement and/or columns of dry hail. In U shapes, dry hail produces little, if any, return. Therefore, this column(s) of dry hail may produce a "notch" shape associated with the storm and in particular associated with the contour. In V-shaped returns, columns of strong vertical movement tend to blow out a section of precipitation, once again forming a "notch."

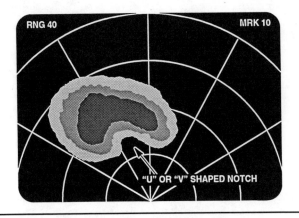

Fig. 3.22 U- or V-shaped notch.

5 and 6. Storms with "hooks" and/or "fingers" are often related to areas of hail and tornadoes.

7. Storms with asymmetric shapes are indicative of steady-state storms (discussed in Chapter 7). Essentially, a thunderstorm matures in an environment where winds change direction and/or velocity with height. As the storm grows it will begin to lean over. The majority of the downdraft is now jettisoned from the top, which is leaning over. Thus, the downdraft no longer interferes with the updraft and the storm can grow and persist.

8. Thin curling shapes in a constant state of change are often associated with tornadoes and other cyclonic winds.

9. A V or oval shape that appears as a thin layer of only rain is highly suggestive of *virga*. Virga can be quite turbulent and should be avoided when possible. It appears as a *thin* layer of

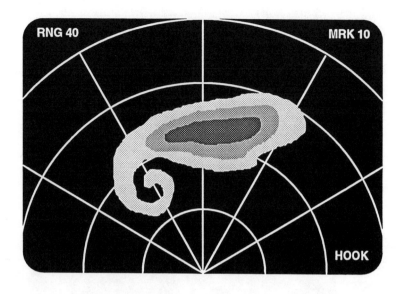

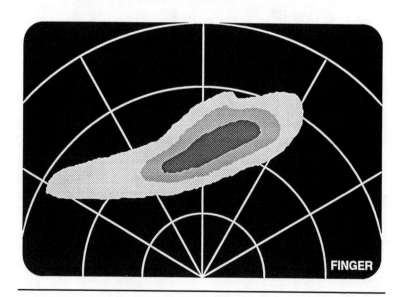

Fig. 3.23 The hook is not the tornado. It is a low-pressure area that draws precipitation into a mesocyclonic circulation. The reflection off the precipitation forms the "hook" echo.

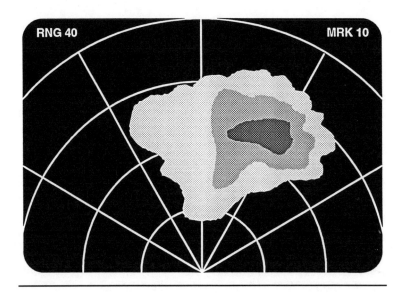

Fig. 3.24 A steady-state storm as it may appear on the PPI. Note the asymmetrically shaped contouring area.

rain. Small movements of the antenna up or down may result in no return at all on the PPI.

10. Small dots depicted in the vicinity of a TRW can be indicative of a hail shaft. Hail shafts are often separated from the main body of the storm. Hail shafts in this form can be quite subtle and difficult to detect. The low reflectivity of the hail and the dimensions of the shaft (normally small, one hundred to several hundred feet or more in diameter) do not offer much reflective material for detection. Delicate tilt management and the short-range scale (for maximum resolution) must be used to detect this illusive hazard. These shafts are frequently located downwind of the storm. Therefore, the shaft is more apt to be detected, and you are more susceptible to it, when you deviate downwind. The hail shaft's distance from the storm is often a function of the winds aloft. Generally, the correlation is: For each knot of wind, the hail can be carried 1 NM downwind. Thus, if the wind velocity is 50 knots, hail could be 50 NM from the storm.

This phenomenon is particularly insidious because (a) it requires an active tilt management program, even though you are already clear of the weather; and (b) the hail shaft return can be very subtle and may go unnoticed. Its appearance is similar to a minute "blip" on the radar screen.

11. Building or broken lines of thunderstorms can mature rapidly, negating safe passage. Although it may be tempting to transition one of these areas—do *not* attempt to do so!

12. Angular targets are normally associated with ground returns. They are included in this section not because they are indicative of hazardous weather but because it is important that they are *not* misinterpreted. Ground returns tend to parallel the range marks and to be more angular than a weather target. Conversely, a weather signature does not normally curve with the range arcs and usually, in fact, crosses range marks. Additionally, as the antenna is tilted up, ground returns should be quick to drop out, whereas weather returns will tend to remain longer.

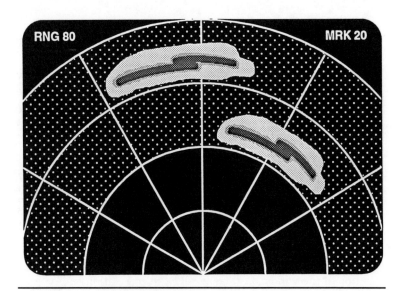

Fig. 3.25 Depicted are prominent ground returns. These returns are angular, parallel to the range marks, and should disappear with modest up-tilt inputs.

Terrain with prominent features will return enough energy to paint red on color radars or contour on monochromatic scopes, if the contour feature is selected.

Proper tilt management and a thorough understanding of the radar tops technique will greatly assist in differentiating ground targets from weather targets. For example, if a target appears that is not easily distinguishable, tilt down so that definite ground returns match the distance of the target. Now use your formula for determining beam width (distance × 100, and change to feet–remember?). This gives you the beam width per degree. If the target is at the 40-NM arc, then each degree is equal to 4,000 feet. Now, if you tilt the antenna up 1 degree or so and the target disappears, you can be fairly certain it was a prominent land feature or city, especially if it possessed the characteristics mentioned earlier. For further confirmation, check your distance measuring equipment (DME) and VOR to determine if a city is in the area in question.

Gradient

The *gradient* is the distance measured inward from the outer edge of the storm to where the storm begins to contour. It is also defined as the measured distance from one rainfall rate to another. In other words, how quickly is the rainfall rate changing with respect to distance? This can be determined by identifying the distance between any two colors if using a color radar, or any two tones if using a monochromatic display.

Steep gradients are areas where the rainfall rate increases rapidly in a short distance. Areas with steep gradients are stronger, more hazardous, and generally better defined. A *shallow gradient* has a longer relative distance from one rainfall rate to the next. Shallow gradients are less well-defined and most likely indicative of less severe weather than a storm with a steep gradient.

We now have two more tools to assist us in accurately identifying the severity of various weather returns (gradient and shape). However, both gradient and, in particular, shape may not

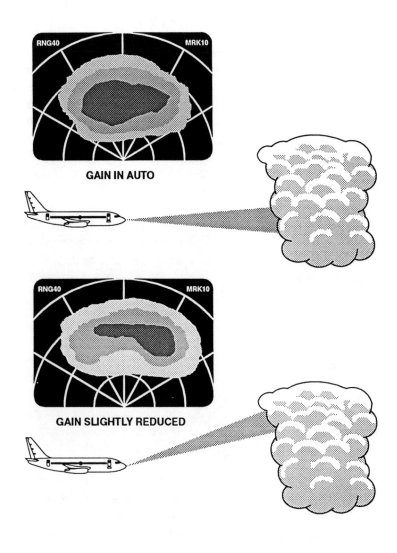

Fig. 3.26 Two different returns of the same target are produced as the tilt and gain controls are adjusted. The lower figure displays an ominous "finger," which only becomes detectable when the antenna is tilted up and the gain is slightly reduced.

readily identify themselves on the radar screen. Therefore, in many cases, the pilot is responsible for uncovering these telltale signs of threatening weather. In most cases this can be accomplished only by incorporating all the techniques discussed thus far. For example, to identify a hook or finger, probing the storm with the antenna (as done when using our radar tops technique), manipulating the gain control, and using the contour function may be necessary.

In this example, the hook or finger could be at any altitude in the storm and only detectable using a probing antenna tilt strategy. Additionally, the hook or finger could be embedded in an area of significant precipitation. It might be identified only if the gain is reduced, revealing the heaviest returns. And finally, the shape may or may not lurk in a contouring area. The pilot may, therefore, need to use both the contour and the weather/normal modes.

Obviously, all aspects of our weather radar are required to accumulate the data necessary to make a well-informed decision regarding weather deviation. Additionally, incorporating all aspects of the radar will enable the pilot to internalize technique and habit patterns that will eventually become instinctive.

Confidence Check

A *confidence check* should always be performed on the radar to ensure proper operation. The most effective confidence check is performed in flight, and, to a lesser degree, a confidence check can be done during taxi out.

While taxiing out, turn the radar on to the weather mode. Do not operate the radar when within approximately 200 feet of other people or highly reflective targets such as aircraft, fuel trucks, or hangars. As you taxi out, tilt the antenna up to detect weather targets. If none are in the immediate vicinity, then select the next longer range and scan for more distant storm systems. If good weather pervades, a confidence check can still be performed. Select the shortest range and tilt down slowly until at least some kind of ground return is indicated. It may not be much of a return, but some return should be displayed. If a city is close by, the returns will possess angular characteristics and be parallel to the range marks.

After take-off a more thorough check can be made. In fact, this check can be incorporated when determining the zero tilt position. When airborne, the confidence check requires that the radar weather mode be selected and the antenna tilt be adjusted to depict *symmetric* ground returns. In some cases, the ground returns may appear asymmetric (lopsided), painting ground on one side but not the other side, or at a different distance on each side. This most likely will occur after rapid acceleration or deceleration, shallow turns, or other maneuvering that spawns precession. If this is the case, the situation should rectify itself in a few minutes. If the asymmetric return persists, the stabilization is inoperative or unreliable and should be addressed.

The confidence check is an important part of weather radar protocol and should be conducted on each flight.

4

Terminal and En Route Weather Evaluations

When approaching the terminal area, you can make an early decision/evaluation of how the weather will impact your flight. If your aircraft is equipped with a *flight management system* (FMS) and an *electronic flight instrument system* (EFIS), this problem is easily resolved. The capabilities of these systems enable you to superimpose the weather onto a display that depicts the airport, the final approach course for the landing runway, and the route to the initial approach point. This gives you an excellent view, early on, as to how the weather will affect your descent, approach, and landing. However, if these systems are not available, another method can be used to make this determination.

If a VOR is on the field, superimpose a direct course to the VOR onto the radar screen. (The azimuth lines will be of assistance in plotting an accurate line.) Now check the *distance measuring equipment* (DME) to the VOR and plot a point of equal distance on the line you have superimposed. The range marks will be useful in plotting this point. This point now represents the airport. If the landing runway is known, draw (superimpose) a second line originating from the airport to represent the final approach course. The picture displayed *at this time* shows you how close the weather is to the airport and the final approach course. However, the picture will not remain real-time current and must be updated regularly as the aircraft approaches the field and/or the weather changes position.

This technique can be embellished by superimposing the missed-approach procedure onto the airport (dot). If orientation is becoming confusing, superimpose the compass right over the dot (airport). If the missed approach calls for a climbing turn to a

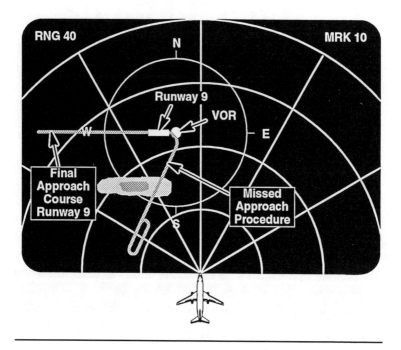

Fig. 4.1 In this figure, the position of the airport, runway 9, the final approach course to runway 9, and the missed-approach procedure have all been established relative to the aircraft and the weather. Superimposing the information on the radar screen provides the pilot with an overall picture of how the weather will impact the procedure. In this case, weather lies on the missed-approach course, necessitating alternate missed-approach instructions.

heading of, say, 200 degrees, draw a line from the dot through the 200-degree mark on your superimposed compass. This represents the missed-approach path. Determine how the weather on the scope will affect it, and advise ATC early if alternate missed-approach plans are in order. As illustrated in Figure 4.1, the published missed approach is unacceptable due to conflicting weather. Therefore, alternate missed-approach instructions should be requested. Although this evaluation may seem premature, it is necessary for your overall situational awareness. In any case, make sure you have a *safe* and *legal* way out by the time you are established on final. If any of these factors concerning your orientation become overwhelming, *you should not be there.* If confusion sets in, don't press on, hoping that things will clear up. *Make a definite decision*; it keeps you in control and your passengers safe.

If the VOR is not on the field or is out of service (OTS), other techniques must be utilized to determine the position of the airport relative to your aircraft. The method considered here uses *point-to-point navigation.* Point-to-point navigation uses bearing and distance to calculate a point relative to your aircraft. The point you are interested in calculating is the position of the airport. A radio magnetic indicator (RMI) is extremely helpful in making this calculation. The computation, though appearing complex, is actually quite simple.

Procedure:

1. The center of the needle or RMI represents the VOR.
2. The tail of the needle represents you (the aircraft).
3. The distance from the center of the RMI to the tail of the needle represents the distance indicated on the DME. (This number is always changing, unless you are flying circles around the VOR.)

The flip side of Jeppensen 11-1 or 10-9 or government equivalent (the plate depicting the airport diagram) will state the airport position in radial and distance form, referenced to a specific VOR. In this example the airport is located on the 090-degree radial 10.0 DME.

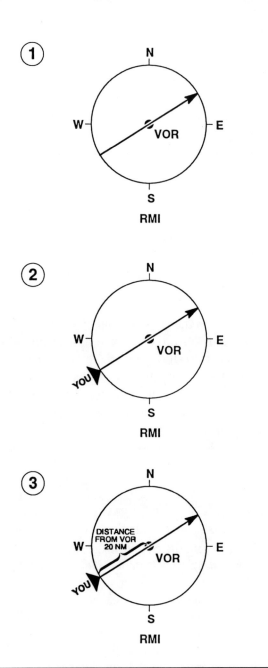

Fig. 4.2 Drawings 1, 2, and 3 in this figure are correlated to steps 1, 2, and 3 of point-to-point navigation in the text.

4. Draw the radial (listed on the plate) from the center of the RMI. (The referenced VOR must be tuned in.)
5. Using the distance stated on the Jeppensen plate, approximate that distance as measured from the center of the RMI. (Remember, you know the distance from the tail of the needle [you] to the center of the RMI [the VOR].)
6. Place a dot at that distance on the freshly drawn radial.

This dot represents the airport. This triangulated picture superimposed on the radar PPI represents a course direct to the VOR, then direct to the airport, with the weather targets oriented to your route. This is the same as taking a standard terminal arrival route (STAR) from your Jeppensen book and overlaying it on the radar screen. In addition, you can also determine the impact of the weather on a direct course to the airport. (If you have not done so already, draw a straight line from the tail of the needle [you] to the dot [airport]. Place a pencil on this line and slide the pencil to the center of the RMI, but don't change the angle of the pencil. The heading resting under the pencil is a heading direct to the airport, not corrected for wind.)

This technique can also be useful for long-range planning concerning weather along your route of flight. For example, in Figure 4.4 you are cleared to VOR A, then are to proceed direct to VOR B, and then flight plan route. You are on a 360-degree heading 50 NM south of VOR A proceeding direct. You see that the airway from VOR A to VOR B uses the 330-degree radial and that the distance between A and B is 70 NM. Using the above technique, you can triangulate a diagram, superimpose it onto the radar screen, and have an excellent orientation to your route and the weather. You, as the pilot, can now make a well-informed and early decision to circumnavigate the storm.

In this case, deviating to the left has several important advantages.

1. You have ample room to give the southwest portion of the line (line of storms) sufficient berth.
2. You are able to deviate *upwind* of the storm system (i.e., deviate around the back side of the line).

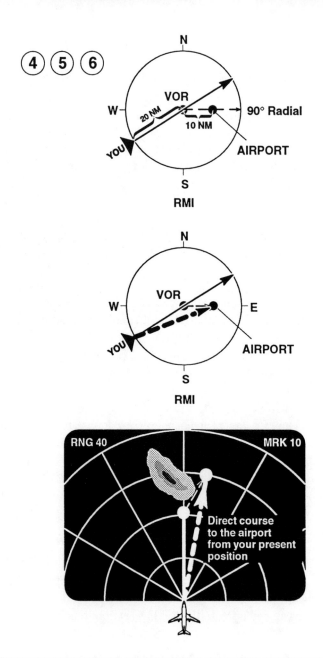

Fig. 4.3 The upper drawing in this figure correlates to steps 4, 5, and 6 of point-to-point navigation in the text. The middle and lower drawings illustrate routing from your present position direct to the airport.

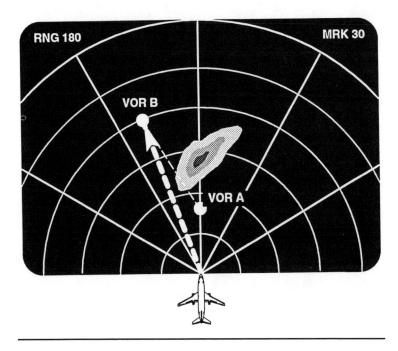

Fig. 4.4 The use of point-to-point navigation enables the pilot to superimpose the route of flight onto the PPI. This provides an overall weather and route orientation.

3. You will have ample opportunity to probe the back side of the line to check for weather. Most likely, this line was attenuating your signal and displaying a *shadow*. Now that shadow can be evaluated.

4. The deviation will probably shorten your route and conserve fuel.

5

Strategy for Weather Avoidance

The culmination of a proficient radar operating technique should result in a strategy that safely guides the aircraft around hazardous weather. This is, of course, the whole objective of the book: How do we analyze the weather with our radar and then make an informed decision on how to circumnavigate it? Let me preface this chapter by emphasizing the importance of a thorough understanding and evaluation of the weather associated with storms. To maintain the continuity of the text, however, the characteristics of storms will be discussed in Chapter 7.

The strategy is divided into two basic segments: (1) the en route horizontal strategy and (2) the arrival and departure strategy.

En Route Horizontal Strategy

After an analysis of a target has been made, a decision whether or not to deviate is in order. If deviation is necessary, particularly close to a target, account for the aircraft's ground track. As illustrated in Figure 5.1, any targets that fall on the

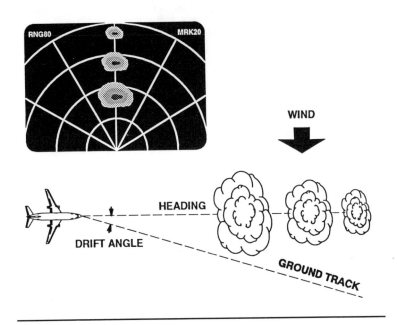

Fig. 5.1 When deviating around weather, consider the effects of the wind, the drift angle, and your ground track. In this case the ground track is working to your advantage. But do not rely exclusively on the heading and ground track differential to keep you clear of the weather—the wind obviously impacts the weather movement.

zero azimuth are targets that lie literally in front of the nose of the aircraft. If you are crabbing into the wind in order to maintain a ground track (i.e., stay on the radial), the weather may not necessarily be located where you think it is.

In this case, it appears the weather lies directly in front of you (and it does); but your ground track will actually clear the weather. (Do not rely solely, however, on the ground track difference to clear the weather.)

A more dangerous scenario results by moving the weather slightly to the right (see Fig. 5.2). Now, according to the PPI, it appears we will stay clear of the weather. But, in fact, the weather lies directly on our ground track.

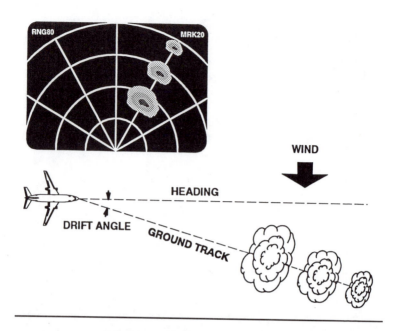

Fig. 5.2 A more dangerous scenario can be developed with a slight modification to Fig. 5.1. On the PPI, it appears that the weather in question lies to the right. However, the harsh reality is that your ground track leads into the heart of the storm.

Therefore, when deviating, always consider the effects of the wind, the drift angle, and your ground track. These considerations are of particular value when avoiding weather that is close to your aircraft.

If a deviation is required, attempt to deviate upwind of the weather. This will normally provide the smoothest ride and reduces the risk of potential hail damage. If the area being circumnavigated is contouring or red, avoid the entire signature by 20 NM (as measured from the outer perimeter of the cell, *not* the center of the storm). Once a storm begins to contour, hail and turbulence may or may not be collocated with the area of maximum reflectivity. If a decision is made to deviate downwind of the target, be aware that *even 40 NM* of distance between you and the storm area may not be enough.

If the winds aloft are strong enough, damaging hail could be carried many miles downwind, certainly in excess of 20 NM. Therefore, if a deviation downwind is necessary, determine the wind velocity at your altitude. Let each knot in velocity represent 1 NM. The total velocity, converted to NM, should be the distance criterion for avoiding the weather in question. Thus, if the wind velocity at your altitude is 50 knots, then avoid the weather area by 50 NM.

Although deviating upwind is the preferred strategy, on occasion conditions may warrant a downwind deviation. For example, in Figure 5.3 a long northeast–southwest line has developed. Your destination lies 70 NM in front of the line. Obviously, an upwind deviation is not practical. However, you could make a significant easterly deviation and approach your destination from

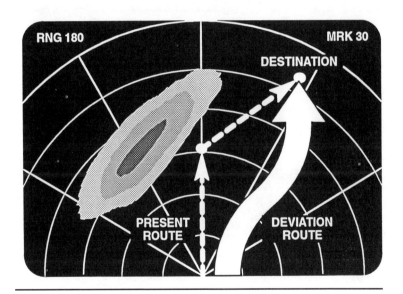

Fig. 5.3 In this particular instance the wind is from the left; should you choose to continue on to your destination, a downwind deviation may be suitable. An early decision to alter your routing to the right should keep you clear of the weather. Additionally, this course alteration is a more direct routing to the destination – thus reducing flight time and conserving fuel.

the south, thus allowing ample separation between your aircraft and the storm.

When approaching a weather target, be sure to use frequent changes in range and tilt in order to avoid a *box canyon/blind alley scenario*. In this case, it appears that the weather in the foreground can be easily sidestepped. However, selecting the next longest range exposes significant weather, forcing the pilot to rethink his strategy.

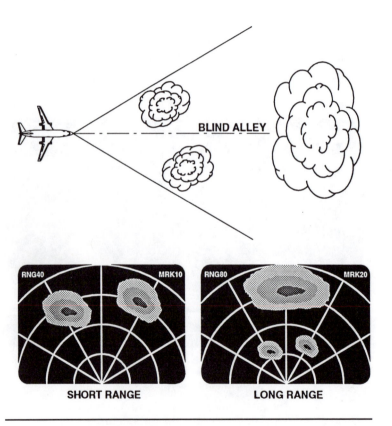

Fig. 5.4 Once again things are not as they seem. The two targets displayed on the short-range scale might suggest a minor side-step maneuver. However, when viewed on the long-range scale, an entirely different picture unfolds. Don't be caught in a box canyon/blind alley situation.

If you are in a rain-producing stratiform cloud while embedded TRWs are in the area, climb. The intervening weather (rain) will add to signal attenuation. If the layer can be topped, a more accurate radar analysis can be obtained while in the clear. Conversely, if you are in the clear (above the stratiform cloud layer) and painting a good picture of the weather, mark its location. As you descend, the intervening weather may degrade the return. If the return is greatly attenuated, at least you will have some idea as to its location. In addition, if the picture is attenuated, run the tilt full up and down in an attempt to pick up the cell tops or bottoms.

Do not rely solely on topping the weather as your only means of avoiding a storm system. There are at least three reasons for this:

1. A reported top is of only historic value and may lack the timeliness necessary for your flight. Even a current pilot report (PIREP) is only an estimate in a localized area. Therefore, even if we assume the report is accurate, it may not contain the scope necessary for your flight.

2. The dynamics of the storm may outperform your aircraft. Therefore, being able to top a line at flight level (FL) 230 does not necessarily ensure safe passage at that altitude 10 minutes later. Severe storms can grow 6,000 to 8,000 feet per minute. A storm building at only 25% of that rate could still prevent many aircraft from crossing the top. Topping a storm by 5,000 to 10,000 feet (although not recommended) may provide a suitable means of traversing the weather. But do not expect a smooth ride. Anticipate turbulence.

3. Be aware of your drift-down altitude. The drift-down altitude is the altitude the aircraft will descend to and maintain should an engine loss occur.

Avoid the flanking clouds southwest of a storm system moving northeast through southeast. If the storm is considered severe, tornadoes are likely to be embedded in the contouring display, but they can be anywhere in the system. These vortices do not have to be positioned vertically but can, in fact, be manifold throughout the storm (see Chapter 7). They can turn horizontal and lurk in the southwest or right-rear flanking clouds of a severe storm. Typically, the tops of the flanking clouds will be from 12,000 to 20,000 feet, with the bottoms at 1,500 to 3,000 feet AGL.

Most important in our airborne weather radar strategy is to use the radar for weather *interpretation, not* weather *penetration.* Think of the radar as a tool used to gauge the distance required to deviate around weather, rather than as a tool used to go through weather.

Arrival and Departure Strategy

The greatest concern when operating around convective activity near the airport is the potential for encountering microbursts, windshear, and severe turbulence. Obviously, during the en route phases of flight, we normally have the luxury of giving all weather that concerns us a wide berth. However, as we approach the airport, both our options and airspace become more limited. When within 40 NM of the airport (and sometimes even farther out), the pilot should have enough information to determine the severity of the weather around the airport of intended landing. The pilot must then determine if this weather will impact his flight in three critical areas:

1. the approach corridor/takeoff alternate
2. the airport area
3. the departure corridor/missed approach

All these elements must be considered simultaneously. A suitable departure corridor must be available should a missed approach occur. If the published missed approach is not advisable because of adverse weather, contact ATC and establish alternate missed-approach instructions. Additionally, on takeoff, not only does the departure corridor require evaluation but so does the approach corridor. If the weather is such that a takeoff is feasible but an approach is not, a takeoff alternate may be required.

Definitive dimensions of the approach, airport, and departure areas are difficult to establish, but, *in general*, a 3- to 5-NM minimum should be maintained between the aircraft flight path and the perimeter of the weather. In some cases the 3-NM avoidance criterion may be more than enough clearance to execute a safe approach and landing. In other instances, 15 NM may not be enough. Conditions that produce thunderstorms and the thunderstorms themselves are so dynamic that other factors must be considered.

Table 5.1 is reproduced from FAA Advisory Circular 00-54, and lists factors and observations associated with windshear probabilities.

Some additional factors to be considered in assessing windshear potential are

- weather movement and velocity,
- whether weather is approaching the airport or departing,
- TRW type (i.e., airmass or steady-state),
- presence of footlike outflow indicative of microburst activity,
- terminal Doppler radar reports,
- ATC-reported field conditions, and
- meteorological conditions responsible for producing the convective weather.

If this information does not seem quantitative enough, it is because meteorology still embraces many unknowns. Therefore, *general* guidelines and recommendations are established to augment the pilot's overall knowledge and experience level.

Each of the factors listed in Table 5.1 is ascribed a probability of windshear. The decision to suspend operations should be based on all these factors on a cumulative basis. For example, if three factors are present and each is correlated with a medium probability of windshear, their collective presence indicates a *high* risk of windshear. Thus, the approach or departure would have to be delayed.

Fortunately, most meteorological phenomena necessitating

Table 5.1. Microburst windshear probability guidelines

Observation	Probability of Windshear
PRESENCE OF CONVECTIVE WEATHER NEAR INTENDED FLIGHT PATH:	
☐ With localized strong winds (Tower reports or observed blowing dust, rings of dust, tornado-like features, etc.)	HIGH
☐ With heavy precipitation (Observed or radar indications of contour, red or attenuation shadow)	HIGH
☐ With rainshower	MEDIUM
☐ With lightning	MEDIUM
☐ With virga	MEDIUM
☐ With moderate or greater turbulence (Reported or radar indications)	MEDIUM
☐ With temperature/dew point spread between 30 and 50 degrees fahrenheit	MEDIUM
ONBOARD WINDSHEAR DETECTION SYSTEM ALERT (Reported or observed)	HIGH
PIREP OF AIRSPEED LOSS OR GAIN:	
☐ 15 knots or greater	HIGH
☐ Less than 15 knots	MEDIUM
LLWAS ALERT/WIND VELOCITY CHANGE:	
☐ 20 knots or greater	HIGH
☐ Less than 20 knots	MEDIUM
FORECAST OF CONVECTIVE WEATHER	LOW

SOURCE: FAA Advisory Circular 00–54.

NOTE: These guidelines apply to operations in the airport vicinity (within 3 miles of the point of takeoff or landing along the intended flight path and below 1000 feet AGL). The clues should be considered cumulative. If more than one is observed, the probability weighting should be increased. The hazard increases with proximity to the convective weather. Weather assessment should be made continuously.

CAUTION: CURRENTLY NO QUANTITATIVE MEANS EXIST FOR DE-TERMINING THE PRESENCE OR INTENSITY OF MICROBURST WINDSHEAR. PILOTS ARE URGED TO EXERCISE CAUTION IN DETERMINING A COURSE OF ACTION.

[This table,] designed specifically for convective weather (thunderstorm, rain-shower, virga), provides a subjective evaluation of various observational clues to aid in making appropriate real time avoidance decisions. The observation weighting is categorized according to the following scale:

HIGH PROBABILITY:
Critical attention needs to be given to this observation. A decision to avoid (e.g. divert or delay) is appropriate.

MEDIUM PROBABILITY:
Consideration should be given to avoiding. Precautions are appropriate.

LOW PROBABILITY:
Consideration should be given to this observation, but a decision to avoid is not generally indicated.

The guidelines in [this table] apply to operations in the airport vicinity (within 3 miles of takeoff or landing along the intended flight path below 1000 feet AGL). Although encountering weather conditions described in [the table] above 1000 feet may be less critical in terms of flight path, such encounters may present other significant weather related risks. Pilots are therefore urged to exercise caution when determining a course of action. Use of [this table] should not replace sound judgement in making avoidance decisions.

suspension of operations will be present in the terminal area for only 15 to 30 minutes. Most delays should not be too much longer. (However, in some instances, airborne holding traffic and/or ground traffic will have built up. Consequently, approaches and departures will require additional time to facilitate sequencing. Some additional fuel burn should be added for your approach requirements.)

If conditions dictate a delay and you have chosen to hold while airborne, several factors should be considered:

1. When calculating your maximum airborne holding time, consider the weather that may be encountered *en route* to your alternate. If weather en route will be a factor, some additional fuel will be needed. Thus, holding time should be reduced accordingly.

2. Select a holding fix where weather will not be a factor. In this scenario, ATC will be quite busy and may instruct you to hold *as published* at some other fix. In some cases, a portion of the holding pattern may skirt the fringes of some weather you would like to avoid. Just moving the holding fix 10 to 20 NM along the same radial may negate your weather conflict. Normally ATC can easily accommodate this request.

3. An ideal holding fix enables you to hold on the back side of the weather or where weather will not be a factor in reaching your destination or alternate.

If convective weather is in the area and any of the other factors mentioned previously are in the terminal area, the pilot must be acutely aware of the risks involved. The pilot must continually reevaluate his decision to continue the approach. At some point it is quite possible to sense a feeling of uneasiness; do not suppress it. Your experiences are instinctively sending up warning flags that must be recognized. Either abandon the approach or *actively* seek out additional information that confirms your decision to continue. For example, request current field conditions. A report of calm winds or no precipitation would be a plus; high gusty winds would be a big minus. Seek PIREP and ride reports. Good news from these reports, however, should be viewed with some caution. Do not rely on good reports or lack of reports for safe passage during periods of dynamic weather conditions. They will not be timely enough. Microbursts are extremely dynamic and occur with a sudden rage. An aircraft just minutes ahead of you encountering a smooth ride is no assurance that you will experience the same. On the other hand, reports of windshear are obvious confirmation that unpredictable weather conditions are present and the approach should be abandoned.

If you continue to proceed with the approach, stack the deck in your favor. Try to:

1. Land on the longest runway.
2. Select a runway that has glide slope information of some sort (either visual or electronic) to assist in immediately recognizing deviation from the recommended glide path.
3. Select a runway that puts as much distance as possible between you and the weather influencing the airport. Some airport runway designs are so spread out that selecting a different runway may put 1 to 2 more miles between you and the weather.
4. Know the criteria indicative of an encounter with, or a precursor to, windshear.
5. Be familiar with the speed additives necessary for this type of approach. Use the recommended speed additives up to a maximum of 20 knots. Recognize, however, the significant increases in landing distances due to minor speed additives.

Thus far we have used the various criteria established to apply our strategies to both approach and departure. While approaching the airport, the options available and our operating parameters tend to narrow. We are, after all, converging on a single point accessible only by the approach corridors offered. To visualize this, imagine the airport situated at the narrow end of a long funnel. As our aircraft approaches the wide end of the funnel, the operating latitude begins to diminish. And as we approach the narrow end of the funnel (near the airport), our available options become extremely limited. Airspeed, altitude, and the ability to continue deviating around weather and still land are greatly restricted. The exact opposite is true during departure. Once again, the airport lies at the narrow end of the funnel, the area where our options and the resources to recover from adverse flight conditions are limited (i.e., extra airspeed, altitude, and power are scarce).

Given the above analogy, it is no wonder that most crashes occur close to or within the airport traffic area. Demands on both the pilot and the aircraft are high. Adverse weather encounters may be enough to exceed the capabilities of either or both (pilot skills and aircraft performance).

Therefore, the guidelines in this section should be viewed as an absolute minimum. They are general in nature and vary with the experience level of the pilot and the capabilities of the aircraft. Remember this: *No* pilot can fly through significant windshear!

Weather Penetration

If for some reason you must penetrate an area of weather, the FAA (Advisory Circular 00-24B) has some recommended guidelines:

Never regard any thunderstorm lightly, even when radar observers report the echoes are of light intensity. Avoiding thunderstorms is the best policy. Following are some do's and don'ts of thunderstorm avoidance:

1. Don't land or take off in the face of an approaching thunderstorm. A sudden gust front of low-level turbulence could cause loss of control.
2. Don't attempt to fly under a thunderstorm even if you can see through to the other side. Turbulence and windshear under the storm could be disastrous.
3. Don't fly without airborne radar into a cloud mass containing scattered embedded thunderstorms. Scattered thunderstorms not embedded usually can be visually circumnavigated.
4. Don't trust the visual appearance to be a reliable indicator of the turbulence inside a thunderstorm.
5. Do avoid by at least 20 miles any thunderstorm identified as severe or giving an intense radar echo. This is especially true under the anvil of a large cumulonimbus.
6. Do circumnavigate the entire area if the area has 6/10 thunderstorm coverage.
7. Do remember that vivid and frequent lightning indi-

cates the probability of a severe thunderstorm.

8. Do regard as extremely hazardous any thunderstorm with tops 35,000 feet or higher whether the top is visually sighted or determined by radar.

If you cannot avoid penetrating a thunderstorm, following are some do's *before* entering the storm:

1. Tighten your safety belt, put on your shoulder harness if you have one, and secure all loose objects.
2. Plan and hold your course to take you through the storm in a minimum time.
3. To avoid the most critical icing, establish a penetration altitude below the freezing level or above the level of $-15\,^\circ$C.
4. Verify that pitot heat is on and turn on carburetor heat or jet engine anti-ice. Icing can be rapid at any altitude and cause almost instantaneous power failure and/or loss of airspeed indication.
5. Establish power settings for turbulence penetration airspeed recommended in your aircraft manual.
6. Turn up cockpit lights to highest intensity to lessen temporary blindness from lightning.
7. If using automatic pilot, disengage altitude hold mode and speed hold mode. The automatic altitude and speed controls will increase maneuvers of the aircraft thus increasing structural stress.
8. If using airborne radar, tilt the antenna up and down occasionally. This will permit you to detect other thunderstorm activity at altitudes other than the one being flown.

Following are some do's and don'ts *during* the thunderstorm penetration:

1. Do keep your eyes on your instruments. Looking outside the cockpit can increase danger of temporary blindness from lightning.
2. Don't change power settings; maintain settings for the recommended turbulence penetration airspeed.
3. Do maintain constant attitude; let the aircraft "ride the waves." Maneuvers in trying to maintain constant altitude increase stress on the aircraft.
4. Don't turn back once you are in the thunderstorm. A straight course through the storm most likely will get you out of the hazards most quickly. In addition, turning maneuvers increase stress on the aircraft.

6

Operational Summary

We have covered a tremendous amount of material. In order to contain the material in an operational format and provide the essence of radar technique, I have devised an acronym. The acronym brings everything discussed thus far together. The acronym is **RAGS**. It can easily be remembered with the following thought in mind: To keep you and your passengers out of rags, use the RAGS index. The index is summarized below:

> **R** radar tops technique
> range
> relative vertical aircraft position
> **A** attenuation
> **G** ground returns
> gain control
> gradient
> **S** shapes
> strategy
> safety

Let us briefly review each item.

R represents the **radar tops technique**. This technique

101

forces the pilot to become operationally proficient in tilt management. All the advantages of this procedure are discussed in detail in the *beam filling* section in Chapter 3.

R also stands for **range**. Any target strong enough to produce a return from 80 NM or more should be considered severe.

R also stands for **relative vertical aircraft position**. Remember, as we take off, climb, cruise, and descend, our antenna will need to be repositioned as our vertical position changes. Antenna tilt and aircraft altitude are inversely related.

A represents **attenuation**. There are various forms of radar signal attenuation. The most important ones are:

1. Radar shadowing: *Never, ever* plan to fly into a target producing a shadow. Additionally, never plan to use a radar shadow area as your only escape route. Until you get around the target creating the shadow, you do not know what lies there.

2. When painting weather while *in* weather, all targets will be *understated.*

3. When painting weather while *in* weather, avoid areas that tend to bulge out toward the aircraft or that appear as thin lines with steep gradients. These are, in fact, the areas of heaviest weather.

4. Sensitivity timing control (STC): Any weather detected beyond 40 NM will be, to some degree, understated.

Recognizing the telltale signs of attenuation is absolutely critical to your survival. As a memory aid, view attenuation as the four S's: *shadowing, saturation* (painting weather while in weather), *shapes,* and *STC.*

G represents **ground returns**. Always paint ground returns somewhere on your radar. An area of rain that produces vivid ground returns behind it will normally be indicative of less-severe weather. You will be more likely to traverse this weather.

G also stands for **gain control**. Use the gain control to identify the most severe weather. As the gain control is reduced, the weaker returns will drop out or begin to fragment. Stronger targets will maintain their shapes. This procedure is particularly useful prior to departure, since ground returns are hard to obtain and zero tilt may not be known. This procedure also lends itself well to navigating through rain with embedded rain showers in the area. Once again, the weaker areas will dissipate while the more significant returns remain.

G also signifies **gradient**. Steep gradients are indicative of rapidly changing rainfall rates and, consequently, severe weather. Shallow gradients are associated with lesser storms.

S stands for **shapes**. Shapes, as well as gradients and contours, can be indicative of types of weather. In general, avoid or be very circumspect of storms that are unusual, irregular, or asymmetric in shape. Such shapes are characteristic of severe weather phenomena.

S also stands for **strategy**. You must decide which way to go and justify that decision.

S also signifies **safety**. Know your procedures and *your own* personal limitations. What may be safe for someone else may not be safe for you.

7

Meteorology

I hope that by this point you have developed an intimate working relationship with your on-board weather radar. It is truly an invaluable ally in the battle with convective weather activity. Since it is imperative to win every battle, it will prove to be most useful if we better understand the enemy: *thunderstorms.*

All weather shares three basic characteristics: (1) *water*, (2) *temperature*, and (3) *stability*. Thunderstorms require one more element, *lifting.*

1. *Water*: All weather is composed of some form of water. The amount of water in the world is constant. The hydrologic cycle consists of sublimation, evaporation, and condensation. Thunderstorms are composed of large amounts of water.

2. *Temperature*: Various ranges of temperature are associated with various weather phenomena. For example, fog requires temperature close to the dew point. Thunderstorms occur when temperatures are in the warmer ranges. The air's capacity to hold moisture increases exponentially with temperature. Therefore, a cold front moving across the country during the summer is likely to produce a number of thunderstorms, whereas the same type of front encountered during winter probably will not produce any thunderstorms.

3. *Stability*: Stability is how a parcel of air reacts to a disturbance. If a parcel of air is stable, it will return to its original posi-

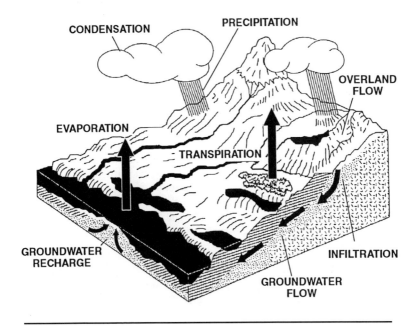

CONDENSATION

PRECIPITATION

OVERLAND FLOW

EVAPORATION

TRANSPIRATION

GROUNDWATER RECHARGE

INFILTRATION

GROUNDWATER FLOW

Fig. 7.1 Like many of Mother Nature's phenomena, the hydrologic process is cyclic. (Adapted from Fletcher Driscoll, *Groundwater and Wells*, Fig. 4.9.)

tion after the disturbance passes. Conversely, unstable air will react violently to even a small disruption. It may accelerate wildly with dynamic oscillations. The meteorologist evaluates the *lifted index* (LI) for stability information and the *K index* to examine the temperature and moisture profile of the environment. This analysis can help determine the stability of the air. Thunderstorms, of course, thrive on unstable airmasses.

4. *Lifting*: All thunderstorms require some type of disturbance to facilitate a violent reaction in an unstable airmass. It may be only the heating of an asphalt parking lot or a much more substantial and dangerous disturbance, such as a rapidly moving cold front.

Moisture must be present and the temperature high enough to accommodate the amount of water required in a thunderstorm.

Additionally, we need unstable air and a mechanism to disrupt the airmass (i.e., thermal or cold front). Although these four elements are all present in *every* thunderstorm, not all thunderstorms are alike. In fact, there are significant differences among them.

There are two basic types of thunderstorms: airmass and frontal [also known as steady-state storms]. Airmass thunderstorms appear to be randomly distributed in unstable air and develop from localized heating at the earth's surface. . . . The heated air rises and cools to form cumulus clouds. As the cumulus stage continues to develop, precipitation forms in higher portions of the cloud and falls. Precipitation signals the beginning of the mature stage and presence of a downdraft. After approximately an hour, the heated updraft creating the thunderstorm is cut off by rainfall. Heat is removed and the thunderstorm dissipates. Many thunderstorms produce an associated cold air gust front as a result of the downflow and outrush of rain-cooled air. These gust fronts are usually very turbulent and can create a serious threat to airplanes during takeoff and approach.

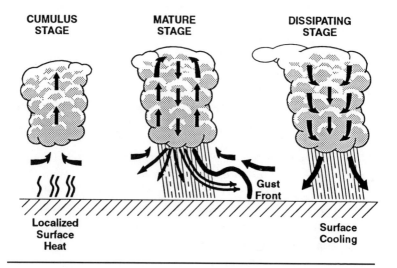

Fig. 7.2 The airmass thunderstorm life cycle has three stages of development: cumulus, mature, and dissipating. Airmass storms are typically spawned by localized terrestrial heating of an unstable airmass. This type of storm is, in general, randomly dispersed. (Adapted from Fig. 2 in FAA Advisory Circular 00-54.)

Frontal thunderstorms are usually associated with weather systems like fronts, converging winds, and troughs aloft. Frontal (steady-state) thunderstorms form in squall lines, last several hours, generate heavy rain and possibly hail, and produce strong gusty winds and possibly tornadoes. The principal distinction in formation of these more *severe* thunderstorms is the presence of large horizontal wind changes (speed and direction) at different altitudes in the thunderstorm. This causes the severe thunderstorm to be vertically tilted. . . . Precipitation falls away from the heated updraft permitting a much longer storm development period. Resulting airflows within the storm accelerate to much higher vertical velocities which ultimately result in higher horizontal wind velocities at the surface.

The downward moving column of air, or downdraft, of a typical thunderstorm is fairly large, about 1 to 5 miles in diameter. Resultant outflows may produce large changes in wind speed.

Though wind changes near the surface occur across an

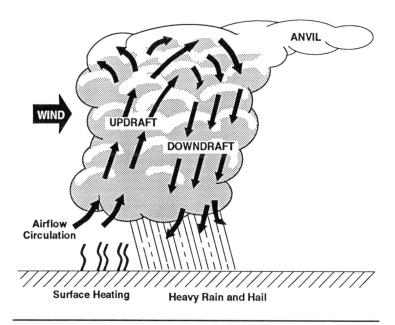

Fig. 7.3 The vertical tilt to steady-state storms enables separate updraft and downdraft shafts to develop. This lengthens the duration of the storm and allows uninterrupted development to occur, consequently intensifying the overall hazardous nature of the storm. (Adapted from Fig. 3 in FAA Advisory Circular 00-54.)

area sufficiently large to lessen the effect, thunderstorms always present a potential hazard to airplanes. Regardless of whether a thunderstorm contains windshear, however, the possibility of heavy rain, hail, extreme turbulence, and tornadoes makes it critical that pilots avoid thunderstorms. [Figure 7.4] shows average annual worldwide thunderstorm days.

Certain areas can readily be seen to have a high potential for windshear because of the high level of convective activity. Due to the lower frequency of air traffic in the highest threat areas (the tropics), fewer accidents have been reported in these regions. (FAA Advisory Circular 00-54)

Now that we are familiar with the types of storms, to what crucial aspects of the weather should we be particularly alert? To assist in answering this question I have designed an acronym: **STAY ALIVE**. If you want to stay alive while operating near convective activity, you had better be familiar with the STAY ALIVE index. Your preflight weather briefing is absolutely critical to the safety of your flight. Even though your radar operating skills may be well refined, you are doing both yourself and your

Fig. 7.4 Average annual thunderstorm days worldwide. The highest threat of windshear is located in the tropical regions. (Reprinted from FAA Advisory Circular 00-54.)

passengers a grave disservice if you do not obtain a thorough briefing. To ensure a comprehensive briefing, use the STAY ALIVE index.

S represents **sigmet**. In this case we are looking specifically for **convective sigmets**. Convective sigmets are issued for any of the following:

1. Severe thunderstorms with surface winds of 50 knots or more and/or hail ¾ inch or more in diameter.
2. Embedded thunderstorms.
3. Line of thunderstorms.
4. Thunderstorms greater than or equal to a video integrator processor (VIP) level 4, affecting 40% or more of an area at least 3,000 square miles.

Whenever you see the words *convective sigmet*, think *severe*. They imply severe turbulence, severe icing, or low-level windshear. These bulletins are updated hourly. The hourly updates consist of an actual observation and/or a forecast. The forecast is valid for two hours after its issuance. When conditions warrant, a special bulletin may be issued at any time. All sigmets will be identified by a number and a letter. The number is a sequenced number. For example, if a sigmet is ongoing, a new sequential number will be issued for each update of the original convective sigmet. The letters following the numbers represent a region in the United States: E = eastern, C = central, and W = western.

An example of a convective sigmet is given below:

MKCC WST 23 1755
CONVECTIVE SIGMET 17 C
KS OK TX
VCNTY GLD-CDS LINE
NO SGFNT TSTMS RPRTD
FACIST TO 1955 Z
LINE TSTSMS DVLPG BY 1855 Z WILL MOV
EWD 30–35 KTS THRU 1955 Z
HAIL TO 1½ IN PSBL

A convective sigmet may be issued under any convective circumstances that the forecaster believes to be hazardous. Remember the term *sigmet* goes hand in hand with the words *significant* and *severe*.

T stands for three items: (1) **tops** (radar tops), (2) **tropopause**, and (3) **temperature** (specifically, the freezing level).

1. **Tops:** If the radar tops are reported to be 30,000 feet or more, consider the storm to be hazardous.
2. **Tropopause:** If a thunderstorm top exceeds the tropopause, the storm is extremely hazardous. The tropopause is the boundary line between the troposphere and the stratosphere. A major characteristic of the tropopause is an abrupt change in the temperature lapse rate. In many cases, temperature may be constant. Although it is unusual for a thunderstorm to exceed the altitude of the tropopause, some do. The tropopause overlying North America is located between 30,000 and 50,000 feet. Therefore, any TRW reaching the tropopause would already be considered severe due to its height.
3. **Temperature:** Specifically, we are concerned with the freezing level. Any TRW whose top exceeds the freezing level by 10,000 feet or more should be considered hazardous.

A stands for **AC**. This is the National Severe Storm Forecast Center's (NSSFC) abbreviation for their convective outlook. AC notes are generated at NSSFC in Kansas City, Missouri, and describe the prospects for general thunderstorm activity for the next 24-hour period. AC notes are also used in preparing the Severe Weather Outlook Chart. Therefore, these AC notes should be associated with severe thunderstorms. The criteria for a severe thunderstorm are

1. surface winds equal to or greater than 50 knots,
2. hail equal to or greater than 3/4 inch in diameter, and
3. tornadoes.

Y represents **your own** observations and intuitive assessments of a situation (which can be quite reliable).

A represents **airmass**. Is the thunderstorm an airmass or a steady-state storm? As previously discussed, there can be a dramatic difference in the severity and longevity of airmass and steady-state thunderstorms. If steady-state storms are involved, avoid the entire radar signature.

L is indicative of **location**. In different regions of the country (and the world), different storms may be depicted on the PPI with identical reflectivity levels. However, there may be a huge difference in their severity. An airmass storm in south Florida is not comparable to the same type of storm in the Midwest. The Midwest storm, for various reasons, will be more severe. Therefore, thunderstorms, like real estate, depend on location.

I is for **information**—that is, other information that might be listed on the terminal sequence reports. These reports can be very comprehensive and include items such as virga, towering cumulus, altocumulus (which is indicative of unstable air), and wind gusts related to rain shower activity. Also included in these reports should be the risk factors discussed earlier, like low-level windshear alerts, temperature/dew point spreads, turbulence reports, PIREP, etc. Obviously, there is much to be considered. This portion of the STAY ALIVE index—information—provides a means to fill in the gaps and round out the picture.

V stands for **velocity**. Normally, if a storm is moving at 20 knots or greater, it should be considered severe. A general rule of thumb to determine the gust factor associated with a TRW is to add 30 knots to the ground speed of the storm. Therefore, 30 knots added to a storm moving at 20 knots would produce a potential 50-knot surface wind. Any storm that produces 50 knots of surface wind is considered severe.

E designates **echo intensity**. Intensities range from level 1 (weak) to level 6 (extreme) storms. These readings are predicated on ground-based meteorological weather radar. The information is then obtained from the VIP and is expressed as a VIP level (e.g., VIP level 3). Echo intensities assigned a VIP level 3 or higher should be considered extremely hazardous. Remember, a level 3 storm is classified as strong and will depict contour on the radar screen (see Table 2.1).

We have covered a wide range of information that will play a useful role in assisting us to better assess the convective weather confronting us. Fortunately, using the STAY ALIVE index provides us with much of the pertinent information necessary for a thorough preflight briefing. Coupling this index with the RAGS index arms us with an arsenal to be deployed as a defensive weapon in circumnavigating severe weather activity.

8

Narrative
Mock Flight

We have now accomplished two most formidable goals. We possess the skills necessary to effectively operate our airborne weather radar (using the RAGS index); and we have constructed a framework to supplement our existing weather briefing (using the STAY ALIVE index). Additionally, each flight segment has been analyzed by using and applying the techniques discussed in the previous chapters. To further solidify and augment the entire process, a realistic narrative of a typical flight during convective activity is provided here.

You are piloting an aircraft from Oklahoma City to Dallas–Fort Worth. It is a hot, mid-August afternoon with high relative humidity. A thorough weather briefing indicates that convective activity will be an influencing factor on this flight. During the cockpit preparations, as before every flight, a good system test of the radar unit is imperative. Unfortunately, on this particular occasion, selecting the test pattern generates an unusual display on the PPI known as *spoking* (see Fig. 8.1).

Maintenance is notified and you refer to your aircraft manual and the manufacturer's radar manual. Maintenance solves the problem by replacing the receiver/transmitter (R/T). Another system test is accomplished, producing the appropriate test pattern. However, you are still skeptical of the reliability of your

113

Fig. 8.1 Various forms of spoking can occur. Quite frequently the fault is in the R/T unit.

radar. Past experiences have revealed that a good system test does not ensure system operation. Therefore, during taxi out (and away from other aircraft and buildings), a confidence check is performed (a confidence check should always be performed on taxi out). As the antenna is slowly tilted down, limited ground returns appear. They are clearly ground returns because they are quite angular and parallel to the range marks. Tower now clears you into position and clears you for takeoff. You request a couple of minutes in position on the runway to evaluate the departure corridor. The tower consequently cancels your takeoff clearance and approves your request.

The antenna is slowly tilted up to probe the departure path using the 40-NM scale. (As you taxied out, you were also evaluating weather returns at various ranges and tilt settings. This is the first opportunity you have to analyze the weather along your takeoff path.) Using as much of the RAGS index as possible, you decide it is safe to continue. The weather returns appear to be just light rain showers, for the following reasons:

R 1. No risk factors are present (virga, lightning, heavy rain, large temperature/dew point spread, etc.).

2. As the antenna is tilted up and down, no contouring or turbulent (magenta) areas are observed.

3. The weather is displayed when using only a narrow range of tilt settings, indicating the weather has little vertical height. (Since this is the first leg with this aircraft, zero tilt information is unavailable.)

A 4. No attenuation is evident; no unusual shapes or intervening weather are present. The current returns are 15 NM away; therefore, you are confident you are receiving an accurate picture of the weather.

G 5. When reducing the gain, the returns quickly fragment and drop out.

S 6. No definite, unusual shapes appear.

7. The active runway is the longest runway.

Given this information, you decide to take off. Flight conditions are smooth and departure is normal.

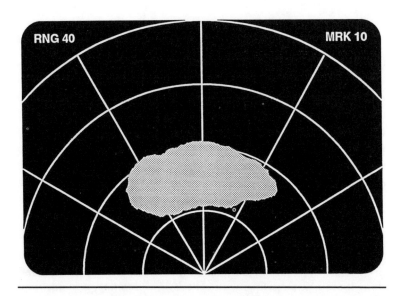

Fig. 8.2 This target appears to be an area of light rain. Use of the RAGS index assists in this determination.

As you climb out, one more confidence check is performed. The antenna is tilted down to obtain symmetric ground returns and compute the zero tilt calibration. Unfortunately, the ground returns are *asymmetrical*. Tilting the antenna down will normally compensate for this problem by producing ground/weather returns where previously none were detected. However, you realize that for the last minute or two you have been in a 5-degree bank in order to slowly deviate around a rain shower as you intercept your airway. Quite possibly, this shallow bank has produced a precession error in the radar stabilization system. The aircraft is rolled wings level and in a couple of minutes symmetric ground returns are observed on the PPI. This resolution indicates that precession error was the culprit; consequently, you decide to use at least 10 degrees of bank whenever you are depending on your radar for reliable returns.

A zero tilt computation can now be made. As you pass through 10,000 feet, with the antenna tilt at zero, ground returns should be observed at 50 NM. This is based on a 4-degree antenna beam:

$$\text{zero tilt} = \frac{\text{altitude}/100}{\tfrac{1}{2} \times \text{beam angle}}$$

In order to produce ground returns at 50 NM, the antenna must be tilted up to +1 degree; this is within acceptable parameters.

As you go instrument meteorological conditions (IMC), you range up on the radar to the 160-NM scale to do some long-range planning. A narrow, short embedded line of storms is depicted at 80 to 90 NM. You request from ATC a top report at your present position. ATC has reports of tops at FL 210. In an effort to regain visual flight rules (VFR) weather conditions, you request FL 250. Even though the flight conditions are favorable at your current altitude, you are concerned that extensive light rain may attenuate the radar signal. As the climb is continued to FL 250, you continually adjust your antenna downward to always paint some ground returns. (Remember this general rule: Aircraft altitude and antenna tilt are inversely related.)

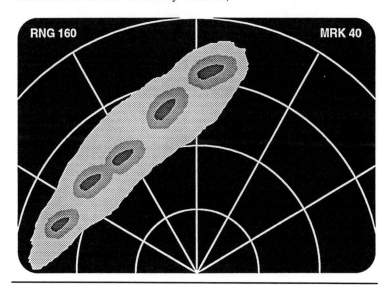

Fig. 8.3 This area of convective weather possesses all of the telltale signs of extremely hazardous weather. The now-building line of thunderstorms could soon merge to form one continuous contouring line. This area should clearly be avoided.

You make an early decision to deviate when you are 50 to 60 miles away from the embedded line of storms. Using the STAY ALIVE index alone, this line should be considered severe for the following reasons:

> Sigmet (convective) was issued for this area.
>
> Tops were reported above 30,000 feet.
>
> AC Kansas City note was issued for the area.
>
> Your intuitive assessment and past experience suggest that this weather is severe.
>
> Airmass TRWs are not indicated; however, steady-state TRWs are.
>
> Location is in a geographical area known for severe weather activity.
>
> Information is available: lightning.
>
> Velocity of the line is 20 knots.
>
> Echo level is 3 or greater.

This particular line of storms meets every criterion of the STAY ALIVE index. But remember, meeting only one element in the index is indicative of severe weather. Further analysis using the RAGS index provides additional confirmation.

> **R** Range was 80 to 90 miles when the line was first observed, providing strong evidence of severe weather.
>
> Radar tops technique depicted a well-defined line, with high reflectivity. Additionally, most parts of the line were contouring.
>
> **A** Attenuation: Shadowing was clearly displayed. Never fly into a return producing a shadow or use the shadow area as your only way out.
>
> **G** Gain: As the gain was reduced, the line maintained its shape, although some parts dropped out.
>
> Ground returns on either side of the line were depicted, indicating clear areas to deviate toward.
>
> Gradients: Steep gradients were depicted.
>
> **S** Shapes were typical of a steady-state line.
>
> Strategy: Evidence for deviation is overwhelming.

Your present position (oriented to the south portion of the line) and the northeast movement of the storm suggest a southerly deviation. Additionally, aircraft to the south of the line have reported good flight conditions. Using point-to-point navigation, you determine a heading that will not only keep you clear of the weather but will intercept a different transition on the standard terminal arrival route (STAR) to your destination airport.

You are now clear of the severe line of TRWs and close to the top of your descent point. The airport is well within your weather radar range and you note several isolated buildups that may be potential hazards during your arrival. As you begin the descent for approach, you determine the position of this weather because you are concerned with possible attenuation as you enter a light rain area below. Dallas–Fort Worth International is landing to the north on runway 36L and 35R. When 40 miles from the field, you determine that the isolated buildups are to the west of the field moving east-northeast. It appears that one buildup is positioning itself near the outer compass locator (LOM) for 35R, your landing runway. (You are operating on the 80/20 scale and will shortly select the 40/10 scale.) To avoid a potential conflict at the LOM for 35R, you request 31R. The runway has an instrument landing system (ILS) but is shorter. You determine that weather will not be a factor with this approach. As the final approach course for 31R is intercepted, you note the weather near the 35R LOM has begun to contour and you *actively* search for other *risk* factors that suggest windshear (i.e., lightning, virga, high-temperature dew point spreads, heavy rain at your position, PIREP, inquiries for field condition reports, etc.). None of the risk factors appear to be present but you decide some additional speed on final would be appropriate. Furthermore, it appears weather could be a factor at the missed-approach holding fix, located some 15 NM from the field. Therefore, a request is made to ATC for alternate missed-approach instructions. They are received and understood and accepted.

At 1,000 feet above the ground, the runway is in sight and you are cleared to land. All of your rigorous planning, evaluation, and conservative strategies are about to pay off as your flight nears completion. As the pilot in command, you experience a feeling of success and fulfillment. Suddenly, ground traffic inadvertently crosses your landing runway. Tower issues an immediate go-around. You are requested to comply with the alternate missed-approach instructions (which, hopefully, you have written down). Soon you will be vectored for another approach, requiring all of the integrated and highly developed skills in airborne weather radar operation. Piece of cake. Right?

Glossary

Airmass Thunderstorms. Thunderstorms randomly distributed in unstable air that develop from localized heating at the earth's surface.

Angle of Incidence. The angle between the radar beam and the target.

Asymmetric Ground Returns. Ground returns that should be displayed across the entire PPI at the same distance but in fact vary in distance; normally displayed as a lopsided return.

Asymmetric Weather Returns. Weather returns depicted on the PPI possessing unusual shapes and/or lacking symmetry.

Attenuation. The loss of signal strength or return.

Azimuth. Direction expressed in terms of a compass using 360 degrees; where east = 90 degrees, south = 180 degrees, west = 270 degrees, and north = 360 degrees.

Azimuth Resolution. See Beam Width Smearing.

Azimuth Scan. The horizontal sweep of the antenna.

Beam Filling. A target's capacity to fill a beam.

Beam Width. The width (diameter) of the beam at any given distance.

Beam Width Smearing. When two separate targets merge on the PPI to appear as only one target.

Blooming. A high-intensity brilliance that flares up and causes the target to lack visual sharpness on the PPI.

Cell. A single thunderstorm.

Confidence Check. A test conducted to ensure proper radar operation and reliability.

Contouring. The area of a storm that returns 40 dBZ or more of energy. This area typically is displayed black on monochromatic and red on color PPIs.

Dew Point. The temperature to which a sample of air must be cooled to reach moisture saturation.

DME (distance measuring equipment). An electronic device used to calculate distance.

Doppler Effect. Explains the rise and fall in pitch (frequency) of approaching and departing objects, respectively.

Drift-Down Altitude. The altitude an aircraft will descend to and maintain should an engine loss occur.

Echo. Normally referred to as the target responsible for reflecting energy or the reflected energy from the target.

Embedded Thunderstorms. Thunderstorms that are obscured from outside visual reference due to intervening weather phenomena.

Erection Torquers. The mechanism that maintains the gyro's vertical axis perpendicular to the surface of the earth.

Excursion Limit. The mechanical and electronic limitations on radar stabilization relative to pitch, roll, and antenna tilt.

Feed Horn. A device located on the main axis of the antenna that carries the microwave energy from the antenna to the splash plate.

Frontal Thunderstorms. See Steady-State Thunderstorms.

Gain. In the context of the text, a control that regulates the listening sensitivity of the receiver.

Gradient. The distance measured inward from the outer edge of the storm to where the storm begins to contour. The change in rainfall rate with respect to distance.

Ground Base. Positioning the antenna to display ground returns at a distance equal to that of the target in question.

Ground Returns. The image depicted on the PPI as microwave energy reflects off the surface of the earth.

Hail Shaft. A corridor or stream of hail, of varying dimension, within or disjoint from an area of convective weather.

K Index. Evaluates the temperature and moisture profile of the environment.

Lapse Rate. The rate of change of an atmospheric variable (usually temperature) with changing height.

Lifted Index (LI). An indicator of stability. It looks at the difference in the temperature of a parcel of air if lifted to 500 millibars and the environmental 500-millibar temperature.

Magnetron. Vacuum tube responsible for boosting the power output in the X band range.

Main Lobe. That part of the antenna beam where most of the energy is transmitted.

Microbursts. Violent downdrafts associated with convective weather, typically a few hundred to 3,000 feet in diameter.

Microwave. An electromagnetic wave with a wavelength of 1 meter or less.

Overscan. A condition where the antenna beam is projected above the weather target.

Paint. The process of detecting a target and depicting it on the PPI.

PIREP. A weather-related pilot report or observation.

Plan Position Indicator (PPI). The radar screen; displays echoes in a plan view using azimuth and range angles to estimate target location.

Point-to-Point Navigation. Triangulates bearing and distance information to calculate a point relative to your aircraft.

Pulse. A single burst (transmission) of electromagnetic energy. The transmission is typically of very brief duration.

Pulse Repetition Frequency. The number of pulses transmitted per second.

Radar Tops. The height of a thunderstorm as determined by meteorological radar. In this text, a technique used that renders a rough approximation of thunderstorm tops but more importantly incorporates many aspects of antenna tilt management.

Reflectivity. A target's capacity to reflect microwaves.

RMI (radio magnetic indicator). Provides bearing information for a VOR. Similar to a nondirectional beacon (NDB).

Sensitivity Timing Control (STC). The mechanism that attempts to compensate for attenuation occurring at short distances.

Shadowing. The absence of signal returns behind a target.

Side Lobe. Energy that is outside the main lobe; normally a small percentage of the main lobe; may be responsible for false targets and/or picture distortion.

Signature. Any depicted weather associated with the predominant storm or storm system.

Splash Plate (subreflector). Distributes microwave energy onto the parabolic antenna.

Spoking. A display on the PPI indicative of an inoperative component in the radar system.

Squall Line. A narrow band or line of thunderstorms.

Stabilization Error. The inability of the antenna stabilization to maintain a constant sweep relative to the horizon.

Steady-State Thunderstorms (frontal thunderstorms). Form in squall lines, are self-perpetuating, last several hours, generate heavy rain/hail, and produce strong, gusty winds. These storms are vertically tilted by large horizontal wind changes (in speed and direction) at different altitudes.

Stratiform. Descriptive of clouds of extensive horizontal development. Characteristic of stable air and therefore composed of small water droplets.

Symmetric Ground Returns. Ground returns displayed across the entire PPI at the same distance.

Thunderstorm. A local storm produced by a cumulonimbus cloud, accompanied by thunder and lightning.

Tilt Management. The process of manipulating the antenna beam to obtain the maximum information about the target.

Tropopause. The transition zone between the troposphere and the stratosphere, characterized by an abrupt change of lapse rate.

Trough Aloft. An elongated area of relatively low atmospheric pressure, most clearly identified as an area of maximum cyclonic curvature of the wind flow—all occurring at high altitude.

TRW. Commonly used abbreviation for any type of thunderstorm.

Underscan. A condition where the antenna beam is projected below the weather target.

Understate. A condition when the full intensity/strength of a storm is *not* depicted on the PPI.

Video Integrator Processor (VIP). A device used to calculate the intensity of precipitation.

Virga. Wisps or streaks of water or ice particles falling from a cloud and evaporating before reaching the ground.

Waveguide. The tubelike conductor that carries the microwave energy between the receiver/transmitter and the antenna.

Weather Returns. The image depicted on the PPI as microwave energy reflects off a weather target.

Windshear. The rate of change in wind velocity and direction with respect to distance.

X Band. The operating frequency range of airborne weather radar.

Zero Tilt. The antenna tilt setting that positions the main axis (center) of the beam parallel to the surface of the earth.

Bibliography

Bendix. 1989. *Pilot's Manual with Radar Operating Guidelines.* Fort Lauderdale, Fla.: Bendix Avionics Division.

_____. 1981. *Weather Radar Operating Techniques.* Fort Lauderdale, Fla.: Bendix Avionics Division.

Bleasdale, Tom. "Weather Radar Now Detects Turbulence." *Air Transport, Avionics,* pp. 18–22.

Boeing Commercial Airplane Company. February 1988. "New-Generation Weather Radar." *Airline Pilot,* pp. 13–22.

Caracena, Fernando. May 1988. "The Microburst." *Airline Pilot,* pp. 17–23.

Caracena, Fernando, Ronald L. Holle, and Charles A. Doswell III. 1990. *Microbursts: A Handbook for Visual Identification.* Washington, D.C.: U.S. Government Printing Office.

Carr, Joseph J. 1989. *Practical Antenna Handbook.* Blue Ridge Summit, Pa.: TAB Books.

Colangelo, Robert V. 1991. *Buyer Be (A)ware: Fundamentals of Environmental Property Assessments.* Dublin, Ohio: National Water Well Association.

Connes, Keith. April 1990. "Bendix/King Vertical Profile Radar." *Professional Pilot,* pp. 56–69.

Cook, Nigel P. 1986. *Microwave Principles and Systems.* Englewood Cliffs, N.J.: Prentice-Hall.

Federal Aviation Administration. 1990. *Airman Information Manual.* Washington, D.C.: U.S. Government Printing Office.

_____. 1989. "Guidelines for Operational Approval of Windshear Training Programs." Advisory Circular 120-50. Washington, D.C.: U.S. Government Printing Office.

_____. 1979. "Low-Level Windshear." Advisory Circular 00-50A. Washington, D.C.: U.S. Government Printing Office.

_____. 1988. "Pilot Windshear Guide." Advisory Circular 00-54. Washington, D.C.: U.S. Government Printing Office.

_____. 1980. "Recommended Radiation Safety Precautions for Ground Operations of Airborne Weather Radar." Advisory Circular 20-68B. Washington, D.C.: U.S. Government Printing Office.

——. 1988. "Rules of Thumb for Avoiding or Minimizing Encounters with Clear Air Turbulence." *Advisory Circular 00-30A*. Washington, D.C.: U.S. Government Printing Office.

——. 1983. "Thunderstorms." *Advisory Circular 00-24B*. Washington, D.C.: U.S. Government Printing Office.

Green, Clifton W. 1985. *Aviation Weather Services*. Federal Aviation Administration and National Weather Service. Washington, D.C.: U.S. Government Printing Office.

Johnson, James P. 1982. *Pilots Handbook of Hazardous Weather*. Schaumburg, Ill.: Aeronautical Training Programs.

Klass, Philip J. 1 May 1989. "Microburst Radar May Spur Review of Tower's Role in Aborting Landings." *Aviation Week and Space Technology*, pp. 79–81.

Kupcis, E.A. May 1988. "Windshear Training Aid." *Airline Pilot*, pp. 29–34.

Lansford, Henry. April 1987. "Avoiding Hazardous Weather." *Airline Pilot*, pp. 18–28.

Lee, J. T. 1964. "Thunderstorm Turbulence and Radar Echoes." *Data Studies, U.S. Weather Bureau*, pp. 17–32.

Levanon, Nadav. 1988. *Radar Principles*. New York: John Wiley and Sons.

Lynn, Paul A. 1987. *Radar Systems*. New York: Van Nostrand Reinhold.

Melvin, W. W. March 1986. "Flying through Microbursts." *Airline Pilot*, pp. 12–17.

——. May 1988. "Landing in Windshear." *Airline Pilot*, pp. 26–28.

Newton, Dennis W. June 1979. "Thunderstorm." *AOPA Pilot*, pp. 41–56.

Ricardi, Leon J. May 1978. "MBA vs. Phased Array: What's Best for Satellite Application?" MSN, 32–37.

Ridenour, Louis N. 1965. *Radar System Engineering*. New York: Dover Publications.

Rinehart, Ronald E. 1991. *RADAR for Meteorologists*. Grand Forks, N.D.: Ronald E. Rinehart.

Schuyler, Norman. November 1985. "Tracking the Elusive Low-Level Windshear." *Aero*, pp. 2–8.

Skolink, Merrill. 1990. *Radar Handbook*. New York: McGraw-Hill.

Smith, Dale. August 1991. "More Reliable Radars Keep Adding Features." *Professional Pilot*, pp. 54–55.

Snyder, Hugh C. 1975. *Aviation Weather*, Federal Aviation Administration and National Weather Service. Washington, D.C.: U.S. Government Printing Office.

Spence, Charles. November 1989. "The Shifting Gusts of Windshear Detection." *Airline Pilot*, pp. 16–18.

Steenblik, Jan W. November 1989. "Windshear Recovery Techniques Compared." *Airline Pilot*, pp. 18–19.

Stroud, Clifton, and Grant McLaren. October 1991. "Westinghouse Radar Uses Doppler Input for CAT and Windshear Knowledge." *Professional Pilot*, pp. 56–58.

Sundarababu, A. 1972. *Fundamentals of Radar*. New York: Asia Publishing House.

Toomay, J. C. 1989. *Radar Principles for the Non-specialist*. New York: Van Nostrand Reinhold.

Towers, Joseph F. January 1989. "Lenticular Clouds: Signs of Turbulence." *Airline Pilot*, pp. 22–23.

Trammel, Archie. March 1981. "Getting a Good Return on Your Investment." *AOPA Pilot*.

———. July 1981. "How to Avoid Thunderstorms Although Radar Equipped." *AOPA Pilot*.

Ward, Neil B., Kenneth E. Wilk, and William C. Herrmann. "WSR-57 Reflectivity Measurements and Hail Observations." WSR-57 Radar Program, U.S. Weather Bureau, pp. 33–40.

Weigel, Edwin P. April 1976. "Lightning: The Underrated Killer." National Oceanic and Atmospheric Administration. Washington, D.C.: U.S. Government Printing Office.

Wiley, John. August 1989. "Windshear on Your Glidepath – What Do You Do?" *Professional Pilot*, pp. 68–71.

0,01-8 µm

Index